COMMUNICATING

BUILDING FOR SAFETY

Communicating Building for Safety

Guidelines for methods of communicating technical information to local builders and householders

Eric Dudley and Ane Haaland

Cambridge Architectural Research Limited

In a survey in Nepal, only three per cent of the respondents identified this drawing as a man running.

Practical Action Publishing
Rugby, Warwickshire, UK
www.practicalactionpublishing.org

First published in 1993
Transferred to digital printing in 2008

Reprinted in the UK, 2018
Reprinted by Practical Action Publishing
Rugby, Warwickshire, UK

ISBN 9781853391835

A catalogue record for this book is available from the British Library.

Since 1974, Practical Action Publishing has published and disseminated books and information in support of international development work throughout the world. Practical Action Publishing is a trading name of Practical Action Publishing Ltd (Company Reg. No. 1159018), the wholly owned publishing company of Practical Action. Practical Action Publishing trades only in support of its parent charity objectives and any profits are covenanted back to Practical Action (Charity Reg. No. 247257, Group VAT Registration No. 880 9924 76).

Document design by Cambridge Architectural Research Limited

A good builder as drawn by a local artist in the Hunza valley in Pakistan.

Contents

When an Urdu version of this cartoon was tested for its usefulness in conveying messages about safer construction in Pakistan, it proved spectacularly unsuccessful.

Introduction

The only form of building improvement programme which has the potential to result in widespread improvements is one which changes the building decisions made by the poor in their own construction projects, designed and paid for by themselves.

- *The* ***task*** *is to encourage practical improvements in indigenous building practices,*
- *The primary* ***tool*** *is communication, and*
- *The main* ***assumption*** *is that outsiders have useful information to communicate.*

The Building for Safety initiative has come about because of certain observations and beliefs:

- Building improvement programmes have had limited impact. Post-disaster reconstruction and pre-disaster mitigation projects often appear to do little for the long-term problems of unsafe housing.
- Even after disasters, most houses are built without the help of engineers or external finance. Aid only reaches a minority of those in need.
- Indigenous solutions are often no longer adequate because:
  - There is a scarcity of important traditional materials like wood
  - Traditional building skills are no longer valued, and
  - There are widespread and rapidly growing aspirations for modern urban-style buildings, without the resources to match.
- The fundamental cause of vulnerability to natural disasters is poverty. There are severe limits to what can be done to mitigate disasters without widespread improvements in the quality of social and economic justice.

Strategies for funding safer buildings have two potential approaches:

- ***Building education, and***
- ***Building subsidy.***

The second option may be of benefit to a lucky few, but for the forseeable future insufficient resources will be available from development agencies for significant subsidies for construction to be widely applicable.

The most sustainable option is when community members co-operate to make better use of the existing resources, often with minor or no inputs from the outside. In this approach, building improvement programmes are intended to act as catalysts through building education projects. They can also help to make people aware of how to use or get access to resources from other institutions. The building education approach is based on two assumptions:

- ***Outsiders can have useful knowledge.*** Development workers possess technical information which communities in the developing world either do not have or do not use, for a variety of reasons. In many circumstances, such information could be used to improve the safety of their houses.
- ***Building education does not require experts.*** The technical principles for making low-cost housing safer are not complicated. They can be learned and disseminated by non-specialist fieldworkers, and can be used directly by householders.

The first of these assumptions needs to be questioned the whole time. Before a development worker can start to be useful in a community, he or she must understand what people already know, how they conceive the need to improve their houses, and the methods they use to do it.

People usually have very good reasons for what they do and for what they do not do. There are many recorded examples of communities which act on accumulated indigenous technical knowledge that has proven to be vastly superior in its context to imported and imposed new ideas. Understanding these practices and the reasons behind them is a key to finding out (together with the local people) what improvements can be made to houses.

Often the problem is not lack of knowledge, but rather lack of resources to put the knowledge into action. In many cultures which traditionally build in stone masonry in areas prone to earthquakes, the sound practice of strengthening the walls with substantial timber ring-beams exists. Yet, today these practices are generally dying out due to lack of timber rather than lack of will and knowledge. In such circumstances a building improvement programme which promotes wooden ring-beams will be redundant. An alternative strategy may be to focus on other types of materials for ring-beams, possibly combined with a reforestation programme. Even where resources do exist, strengthening buildings may not be seen as a priority. Paying for medicine, school fees, or chairs for a guest room may take precedence over making one's house stronger.

In other cases the study of indigenous technical knowledge may reveal that people have ideas which they consider sound but which from a technical perspective do not work. Here, a dialogue between the technical development worker and the community, based on an understanding of the reasons for the community members' practice, might initiate a process of change. This is provided that the new solution is economically, socially and culturally acceptable.

The issue is not to judge people for what they are doing, but accept that they have good reasons for doing it, and that change will only come about when the reasons for changing are better – from their perspective – than the reasons they have for keeping their old practice.

This does not mean the development worker has to agree with what he or she sees and hears. However, it means that an attitude of respect for other people's practices is a prerequisite for gaining understanding and for engaging in a constructive dialogue about possible alternative techniques. An attitude of judging current practices as wrong prevents the development of such an understanding.

The second assumption, that building education does not require experts, is a challenge to the engineer's mystique and expertise. The non-specialist fieldworker can in many cases do a useful job in the community, putting life-saving knowledge into the hands of those who need it. To be able to do this, the fieldworker needs to learn basic technical ground rules and good communication skills which enable him or her to understand about local building practices and traditions before starting to give advice. Only then will the advice have the potential for being practical and relevant to the local house owner. Only then will the fieldworker have the potential for being respected by the community as a person whose advice is worth listening to.

Who is this book for?

This book is written for people who are, or in the future may be, involved in the planning, implementation and evaluation of building education projects. Specifically, people who:

- Are occasionally called upon to assist communities after a disaster
- Regularly discuss and plan building improvement programmes
- Assess problems and possibilities for change in building practices
- Implement building improvement programmes in the field
- Design educational materials
- Direct and supervise those who design educational materials
- Co-operate with trainers and builders who use educational materials, and
- Evaluate building improvement programmes which make use of educational materials.

Some of these people are already working in building education on a regular basis, in governments or NGOs (non-governmental organizations). Others are working in organizations which are called in to assist after a disaster. But often the largest burden falls on to organizations and individuals who are already working in an area before a disaster has struck and who, though having no specialist knowledge of building improvement, suddenly find themselves faced with a massive task of reconstruction. Above all, this document is intended to help such people avoid the pitfalls which many others have experienced before.

What is this book for?

Planners and implementors of building improvement programmes are invited to consider the role and importance of communication at all stages, not just in the development of educational materials, which may or may not be necessary. This document outlines:

- Experiences from projects which have used these approaches
- The principles that have been drawn from these experiences, and
- Some basic guidelines for information gathering.

For the planners and designers of educational materials, we set out practical guidelines on how to develop educational materials on building improvement issues. We start with the stages one must go through to find the right messages to communicate and to decide whether or not to produce educational materials. These stages are often ignored, especially when projects are pressed for time and materials must be produced quickly. The result is often a waste of time and money, and lack of impact on the target audience.

The major emphasis is placed on developing printed materials, with illustrations. There are two reasons for this. First, building education programmes seem to concentrate on this method of communication; and second, there is a body of useful and solid experience relating to graphical material which has relevance for all communication media. There is no detailed discussion of other ways of communicating, which in some programmes may be more appropriate than printed materials or which may be used in combination with such materials. These include techniques such as using traditional story-tellers, local singers, local theatre or puppet shows, video, advertising in the local cinema, and demonstration buildings.

This document does not pretend to prescribe universal solutions. Rather, the intention is to assist the local process of planning and design by highlighting the kinds of questions which need to be asked. In this process, the document can be used in several ways, including:

- To stimulate thinking about communication
- To indicate what methods of communication can be appropriate
- To identify when certain steps in the planning have not been followed, and so highlight potential problems with educational materials and strategies, and
- As a direct tool for developing educational materials.

Digging the foundations for a house. A local artist in the Hunza valley in northern Pakistan was asked to illustrate this theme. In this drawing he has represented the house and the idea of foundations by drawing a two-dimensional plan linked to his three-dimensional image of a man working.

1. Communication in planning

Communication between the development institution and the community is a prerequisite for a project to arrive at a correct definition of the problem, and realistic and sustainable solutions which people will adopt. The starting point for designing a building education project should be to understand how people learn rather than how to teach.

Developing the concept for a building education project is a creative process based on:

- Continuing dialogue between planners and beneficiaries
- Planners understanding the needs, aspirations, and the economic, social and cultural limitations and possibilities of the beneficiaries
- Common, realistic goals, and
- Agreement on roles and responsibilities between the planners, the implementers and the community.

Community participation

A building improvement programme will only be successful if what it has to offer is acceptable to its target audiences. Identify your target audiences and design your intervention on the basis of an understanding of their needs, aspirations, and resources.

There are various levels at which participation can take place – they can be regarded as steps on a ladder, where the level of control over decision-making and implementation in the community is increased with each step. The success of a programme is heavily influenced by the level of involvement of the community, the way such involvement is sought, and on the experience of the development workers.

The lowest level of participation is **compliance** – the development agency seeks the approval of the community to plans already made. The next level is **contribution,** where the community is invited to comment on the plans, and suggest changes. The third level is **collaboration**, where there is an exchange of views before plans are made, and the community and the development organization are equal partners in the process. The final level is **control,** where the community is given the full responsibility for decision-making and implementation with the development organization acting as advisers.

Even though the need for community control has been recognized by development planners for more than a decade, it is not often employed in practice. The two lowest steps on the ladder – compliance and contribution – are the most commonly used strategies. Where true collaboration and control exist there is a greater likelihood of long-term success. But realizing the aim of true community control is not always easy. However well intentioned the assisting organization, it will often be faced with three problems in implementing full community participation:

- ***Who is the community?*** Other than in limited and so-called model projects it is not possible for a project to consult with all members of the community and to meet the expressed and sometimes conflicting needs of all of them. The consultation process can only be with a representative sample of the community. It is a parallel to the market research of commercial enterprises. Sometimes the opinions of those most in need, the marginalized and oppressed, are not readily determined

- ***Inability to express needs.*** People rarely have a fully articulated understanding of their own situation. Many people are not used to the kinds of what if questions necessary when contemplating the various aid options which might be available in the future. Not surprisingly, prospects of aid which involve direct tangible material benefit to the individual are generally more immediately attractive than aid in the form of technical information. In the wake of a disaster, this kind of reflective analysis of one's problems is even more difficult with opinions distorted by immediate memories. The best problem definition will emerge from a dialogue initiated well before the disaster.
- ***External, but legitimate, criteria.*** Development institutions have their own legitimate criteria which may be at odds with those of the affected population. For instance, demands for wood and land may conflict with longer term ecological considerations which an institution cannot responsibly ignore. Or the population affected by a disaster may be relatively prosperous and the kind of aid which they demand may be out of balance with that available to impoverished people elsewhere in the country who are obliged to cope with the slow disaster of day-to-day life.

Apart from the community of householders on the ground, there are also communities of government officials, development fieldworkers, and others who each have their own characteristics and desires. Early in the planning process a project planner must identify:

- ***Primary audience.*** Who is the primary target audience which is expected to implement the new ideas, for example householders?
- ***Secondary audiences.*** Who are the secondary target audiences who could influence the primary target audience to consider or adopt the new ideas, such as local political or other leaders, influential farmers, builders, local shopkeepers, and so on? Such respected people can help lend credibility to the idea.
- ***Facilitators.*** Who are the target audiences on a regional or national level whose support may be needed for the local efforts? They may, for instance, need to help pave the way with organizations or government to facilitate access to resources and information.

Each of the target audiences may need different kinds of educational inputs, through different media or channels. The ideal project design involves all target audiences in the planning of the project. In such a process, the communication of the ideas becomes a much easier task. Each target audience will have its own members in the planning process, and these members will facilitate communication with their group.

Community participation does not guarantee that good and achievable ideas will result, or that a successful programme will follow. It is no substitute for good ideas. However, it is a means of checking the validity of an idea and creating a context in which relevant and useful ideas can emerge.

Whatever problems may be encountered in implementing community participation, the logic of a process based on a continuing dialogue with the project beneficiaries and all the other actors in the process demands project proposals which are highly flexible. There is no point consulting with the people unless their views can feed back and actually change the course of the project.

Information gathering in the field

Make time to listen to people. Before an interview make sure that the respondent understands who you are and why you want the information. Use open-ended questions which allow people to express their real views.

A thorough assessment of conditions in the community in which one is going to work is a necessary early step in the planning process. The dialogue with the community can help to establish:

- ***Problem definition.*** The nature of the problem, both from the perspective of the community and the technical workers
- ***Technical appropriateness.*** Why proposed ideas and solutions are or are not technically appropriate
- ***Cultural appropriateness.*** Whether the ideas are appropriate in terms of local values and resources
- ***Present practice.*** People's present practices regarding the ideas, their attitudes and their knowledge, and
- ***Constraints.*** Reasons why people might not want to participate in the project or adopt a given idea.

More specifically, a building improvement programme may need information about issues such as:

- ***Housing quality.*** The quality of the present houses, and the householders' perception of the quality. Where are the weak spots? Why are these weak? Under what circumstances are they dangerous? What do people do about this now? What do people think can be done to strengthen the houses against natural hazards? Is anyone in the community doing this? Who is he or she, and why is the person strengthening their house?
- ***Hazard awareness.*** Approximately how often do natural hazards occur in the community? What kind of hazards? When did the last one happen? What kind of damage did it cause? What do people do after a hazard? Is there any organized community action to repair the damage, or is this completely the responsibility of individual householders? Do people have the resources, and the willingness to prioritize the use of them, for strengthening the weak points of buildings after a disaster?
- ***Prevention awareness.*** What is the perception of prevention? Is there a tendency to do something to prevent damage from hazards, or are disasters so rare that people would rather resort to repair afterwards?
- ***Sources of inspiration.*** When people build new houses, where do they get their ideas for making changes in traditional building practices from? What are their aspirations? What kind of house would they really like to have?
- ***Community buildings.*** Are there any community buildings which have made use of improved building practices? Has anyone in the community copied these techniques? If yes, who and why, and if no, why not?
- ***Domestic economy.*** How do people spend their money? What are their highest priorities, and why?
- ***Sources of advice.*** To whom do the householders in the community go for advice when they consider improvements to their houses? Who are the so-called master builders who carry on the old building traditions? Who are the younger, more modern builders who bring new urban techniques to the community? Where and in what

circumstances do people discuss building issues?

- *Communication media*. What kinds of media – electronic, printed, and traditional – are available in the community? Do people have radios? If yes, what programmes do they listen to? Who listens and when? For instance, if you want to give men information by radio, you have to establish the most likely times they will listen. Do people have televisions or videos? Is there a community television? Are newspapers available? Who reads them? Are posters, pamphlets or other promotional materials used by any development programme? Do development workers use aids such as flip-charts or slides if they give an educational presentation? Are printed materials displayed anywhere in the community? If yes, where, and on what subjects?

This is not a complete list of questions. It is simply indicative of the kinds of information an organization needs to gather before starting to design a building improvement programme. To gather such information poses a number of challenges in terms of methodology, as well as in terms of implementation. How do you gather information? Which skills are important?

Information can be gathered through informal discussions with householders and groups of community members. This is usually referred to as qualitative information gathering. Alternatively, one can gather quantitative information through questionnaires where the same questions are administered to a certain number of householders in the community.

The questionnaire method has been the most commonly used, and abused. Problems often occur when interviewers are people, such as urban university students, insufficiently trained in how to establish contact and build credibility with their respondents. Sometimes interviewers have a lack of familiarity with and respect for rural people. The use of questionnaires with set options for answers, such as yes-no, or never-sometimes-often-always, is often favoured by planners, as these are quick to analyze, especially with the help of computers. However, the information gathered in this way usually does not give a good picture of the situation, and in many cases is misleading.

To gather the kind of information suggested previously, a qualitative approach will usually give the best results, with interviewers conducting discussions with individual householders and groups of community members on a number of issues. The skills and attitudes of the person collecting the information is the main factor influencing the results of such social research. Interviewers should receive good training in: preparation; building confidence; interview technique.

(a) Preparation

The interviewer, in collaboration with the project planners, should decide in advance what information is needed. A great deal of the interviewer's and the respondent's time is often wasted gathering information that has no practical use, but someone has decided it would be nice to know about.

Part of the preparation is deciding what kind of people you want to interview, and how many. As an example, it would be appropriate in a community of 1000 people to interview 15 to 20 individual householders and two or three groups of four to seven householders each. Each individual discussion would take one to two hours, and group discussions at least two hours each. Some of the group, as well as the individual, interviews should be with people who are sought out for advice on building issues by other community members.

(b) Building confidence

Establishing a comfortable social setting where the respondent feels at ease is the first step in the process. This could be in or outside his or her house, where you are reasonably undisturbed from interruptions. The presence of other people during the interview may influence the results.

Then come the most important tasks. Take time to discuss the purpose of your job. Make the respondent feel you are not testing their intelligence but that you are genuinely interested in his or her opinions. Tell the respondent how the results of the survey will be used. He or she needs to see clearly that there is a good reason to answer your questions conscientiously. If you fail in this task, the information you gather will probably be useless and in some cases harmful, because when planners use such information to design projects, the results can be disastrous.

How do you know when you are receiving wrong information? Most rural people are usually patient and polite. They may give you time and tell you what they think you want to hear, rather than telling you that they do not understand or believe in your task, that you have behaved in an unacceptable way, or that the question does not make sense. It is up to the interviewer to prevent this from happening, to develop antennae to discover when it is happening, and then to be able to take necessary action to correct the mistakes. This usually requires practise. Learning will be quicker if you can get feedback on your techniques and behaviour from an experienced interviewer observing you.

Who you are in the community will influence the answers you will get. If the interviewer is a fieldworker living in the community, the chances are he or she will already have built up credibility with the householders. On the other hand, a known representative of an aid-giving organization could create expectations which colour the responses. An urban university student will usually have more difficulties building up confidence before starting the interview, and will have to devote more time to it.

(c) The Interview technique

It is usually best to work in pairs, one person doing the talking and the other recording what is being said. During the interview there are some basic **dos** and **dont's** the interviewer needs to observe. The interviewer should try to:

- ***Use open-ended questions.*** Open-ended questions are those which cannot be answered with a simple yes or no, or some other predefined answer. They ask people to describe what they think or feel. For example, what are the weak points in your house during an earthquake? How do you decide what changes to make when you repair your house after an earthquake? What are the reasons people here build their houses in this particular style? Such questions are not so easy to ask, the answers can be long and difficult to record, and they usually require follow-up questions to clarify what the respondent thinks. Asking open-ended questions requires the interviewer to be skilled and patient.

- ***Encourage people to talk.*** Listen actively and give full attention to the respondent. Allow him or her time. Give non-committal feedback like 'interesting', 'please tell me more about it', and also non-verbal encouragement.

- ***Be neutral.*** The task is to make the respondent feel that whatever he or she says is OK, and that it is valuable and interesting.

- ***Approach sensitive issues with care.*** If it is necessary to ask about sensitive issues, such as money, start with neutral questions. Only move on to the sensitive subject if and when trust and rapport have been established. In these situations it is particularly important that the social setting is safe for the respondent, and that it has been made clear how the information will be used. If people do not wish to discuss the issue, the interviewer should not insist.
- ***Respect people's time.***

The **don'ts** for the interviewer are:

- ***Do not use closed or leading questions.*** For instance, are the corners in your house the weakest spots during an earthquake? Are the traditional houses cold during the winter? Closed questions are easy to record answers to, and easy to analyze. However, the disadvantage is that the question leads the respondent to answer in a certain way, and it does not invite further discussion or comments that could give answer to the question, why is it like that? A leading question will usually confirm the idea of the interviewer rather than seek out the ideas of the respondent.
- ***Do not be seen to judge.*** If the respondent feels that the interviewer is judging the information as correct or incorrect, stupid or clever, it is likely to colour his or her responses and attitude. Many development workers have experienced that community members quickly sense a judging or paternalistic attitude, even if it is not verbalized. Non-verbal communication is responsible for a substantial amount of the signals we send to another person. The attitude that betrays lack of respect for local people's traditions and practices is probably the cause of the most severe communication gap between development workers and local people. The interviewer's perceived reactions may be unintentional. Without care, the friendly smile can be interpreted as laughing at the respondent's poverty. The task is to find out what the respondent thinks, and hide one's own feelings and reactions to what is being said, to avoid influencing the respondent in any possible way.
- ***Do not argue with your respondent.*** Do not argue or contradict, or the respondent will soon close up or else engage in an exchange of views, which is interesting but does not serve the purpose of information gathering.
- ***Do not try to teach.*** If the interviewer teaches, the respondent will be confused about the purpose of the task: does the interviewer want to hear his opinion, or give advice? The two cannot be combined. If the respondent directly asks for advice, the interviewer should suggest that this be done after the interview or else refer the respondent to someone locally who can give advice.

One of the most important purposes of interviewing members of the community is to try and understand the process through which they acquire new ideas and values. Before one can design an effective communication strategy one needs to understand how and, from whom, people acquire information. The starting point for designing a building education project should be establishing how people learn rather than how to teach.

How people change

Different types of information for different moments in the process of change demand different channels of communication. Find out how people communicate and acquire new knowledge in their daily lives.

In the process of moving from ignorance of a technology or idea to trying it out and deciding to use it in everyday life, a person requires:

- Different kinds of information
- Through a variety of sources, and
- At different moments in the process.

Although the nature of the information and sources varies enormously from place to place there is a universal process of adopting new ideas. This can be described in five main stages:

- ***Awareness***
- ***Interest***
- ***Trial***
- ***Evaluation***
- ***Adoption.***

The first two stages deal with the idea – what is the new technology or idea, and why should the person consider using it? This is the intellectual process of making a choice to try out something new. It does not involve action.

The next three stages deal with converting the idea into practice. Based on the information about what and why, the person makes a decision about whether or not to try out the idea. If the person decides to try it out, he or she will then assess whether or not the idea or technology is suitable and then make the final decision whether or not to adopt the idea or technology. The idea can be rejected at any stage in the process. Most ideas we are confronted with never get beyond the awareness and interest stages.

This model of the change process has important implications for building improvement programmes.

Why-to information. First, people need information about the why-to aspects of the idea and time to consider it. These aspects must be introduced in a manner the person experiences as relevant to their needs. The why-to information relates to the first two stages, awareness and interest.

How-to information. When people have had time to consider reasons why they might try a new idea or technology, and decided they will try it out, only then will they be ready to listen to how-to information on putting the idea into action. This relates to the last three stages in the process.

Many projects make the mistake of starting their education with how-to information. Usually, this does not work. People will not begin at the third stage in the process of changing behaviour, especially when the behaviour is connected with traditional values.

For a programme to reach people with the right information at the right stages in the decision process, it must consider what channels of communication are most effective for each stage. During the first two stages people will be open to information from almost any source: radio; newspapers or other mass media; people they know and respect such as family, community leaders, religious leaders, builders or other craftspeople; development workers, such as extension workers in agriculture or health; or anybody else. Most householders will be ready to consider new ideas, and tell you theirs. Whether or not they take the ideas to the next stages (the action or practical stages) depends on:

- How good the why-to information has been, that is, how relevant it has been to the householder's economic, social, and cultural needs, and
- Who is communicating the information.

The householder is most likely to decide to take action if the idea is communicated to him or her by a person he or she respects or aspires to emulate. This person could just use discussion (the most common method). In addition, printed or other materials describing how to implement the new idea may be employed as a support in the discussion with the householder. However, such material alone, whether distributed to householders in booklet or poster form, or broadcast to them through radio, will usually not be sufficient to make them decide to try it out.

The same is true for the next two stages, evaluation and adoption. This is when the householder evaluates whether the new idea works for him or her, and decides to make it a permanent part of their life, or rejects it. Discussion with individuals he or she feels close to and respects is the most important influence in these stages. As for the trial stage, information through mass media and printed materials can be a support, but by themselves will not be enough to make him or her decide.

Understanding the stages of behaviour change has obvious implications for development programmes, and for the way these development programmes plan their educational interventions. It influences when and how and through whom to communicate. A commitment to a community-based planning process does not mean you have to carry out a two-year research programme before lifting a finger or opening your mouth. But it does mean you have to spend some time discussing ideas with local people, and also find out what people do, what they do not do, what they could or would do, and the reasons they have for their choices. A close contact with the field, through sensitive communication methods with the right people, is the prerequisite for knowing what to communicate.

Summary: **Communication in planning**

- ***Respect local knowledge and aspirations.***

  A successful building education project must build upon existing local knowledge and help to satisfy the aspirations of the target audience. To determine the knowledge and aspirations requires a dialogue based on mutual respect

- ***Involve the beneficiaries at all stages.***

  Communication skills are required at all stages of a project and not just for the dissemination of how-to knowledge, and

- ***Before trying to teach, find out how people learn.***

  A building improvement project intended to introduce new building practices can best start by determining how new practices are already entering society.

Case study: Taxi drivers in Peru

In Peru, the Intermediate Technology Development Group is involved in the Alto Mayo reconstruction plan in the wake of an earthquake in 1990. It has realized that the existing social structure has its own powerful mechanisms for the dissemination of information and opinion. In the September 1992 edition of Appropriate Technology, *which was devoted to communication issues, Andrew Maskrey and Julia Vicuna reflected on the place of taxi drivers in the informal communication network.*

Taxi drivers play a very important role in the communications of a community. This is because they have such a large influence on the public who use them, and because they are considered to be indicators of what the public thinks and wants.

As a rule, taxi drivers are talkative, sparking up conversations with their passengers before they are even through the door, and they have an opinion on every topic of conversation – politics, weather, sport, culture, fashion, and so on.

Most taxis in Peru are not in very good repair. They are old and uncomfortable cars. But passengers do not usually notice this as they are engrossed with the agreeable conversation offered by their drivers.

Before opinion surveys were developed in Peru, popular feeling was measured by polling taxi drivers who were the barometer of public opinion. Although surveys are very scientific these days, taxi drivers are still the public's spokespeople.

Taxi drivers are an effective means of circulating person-to-person information. While they create and express public opinion in big cities, in small towns they know intimately all that is going on.

When IT's (Intermediate Technology, Peru) engineers, who had already been working in the region before the earthquake and many of whom were from the region itself, went to Alto Mayo after the 1990 earthquake, the first people who knew they were coming were the taxi drivers. Then, they made themselves responsible for spreading the news.

In order to evaluate the damage, the engineers had to go to many towns to gather data and interview people. The technical experts travelled by taxi, a much-used service in the region as there are no buses.

The earthquake and its effects, the reconstruction, and the housing problems were all topics for conversation. The taxi drivers were interested in the work that IT was doing in the region. They wanted to know exactly what was going on so that they could discuss it later.

'Engineers from Lima have come to work on reconstructing the houses. They say they are going to use *quincha mejorada*, and it sounds a very interesting project', they said.

Thus, the taxi drivers performed a very important role in the dissemination of the Alto Mayo Reconstruction Plan to the people who rode in their taxis. They introduced the plan, explained what it was all about, where it was happening, and how many people were benefiting from it.

They also talked about other IT activities in the area. People came to look at the projects having heard about them from a direct and known source – the taxi drivers.

2. Educational materials

Concentrate on one or two essential messages. Identify clear targets and educational contexts and use a mix of mutually reinforcing media. Where possible use the real thing. Educational material is only as good as the staff that are using it. See staff are well trained and understand the problems and opportunities of different educational materials. Never assume. Test everything.

After the project has collected the necessary information with and from the community, and arrived at ideas and technical solutions which are economically, socially and culturally feasible and acceptable, it is then time to consider how to communicate the ideas most effectively to the various target audiences.

The principal uses of educational materials in projects are:

- ***To make noise.*** Creating or raising awareness of the existence of a project, and stimulating interest
- ***As a memory aid.*** Reminding people of something which they have already learned, and
- ***As a teaching tool.*** In a school or training course.

Communicating a message may involve the use of educational materials. But such materials will rarely be enough on their own. The skills and attitudes of the fieldworkers using the materials with the target audience are even more important than the quality of the materials. A person with a respectful attitude and good communication skills can turn a poorly developed piece of material into a reasonably good communication tool. A fieldworker without such skills can use a beautifully designed piece of material and lose people's interest.

When something is explained well, people can interpret complicated images about it and understand complex messages. In most cases, good explanations are not available. And even with good explanations, a well-drawn and comprehensible picture is a much better communication tool than a poorly drawn one. The first leaves the respondent to place all his or her energy on learning a new idea. The second demands that he or she struggles with the interpretation of a bad picture.

In the same way, there is no substitute for a good fieldworker. Good communication skills are no replacement for a good idea. People will only take up new ideas when they believe that what is being communicated will work, and that there are good reasons from their own perspective to invest in the new ideas.

Choice of media

Choosing the right channels for communication should be based on a careful consideration of the nature of the message and the characteristics of each medium of communication. Usually, a combination of different media to reach the same audience with the same message will be the most effective.

The choice of how to communicate the ideas, and through which channels or media, should be guided by an analysis of the following issues:

- ***Who is the target audience?*** What is its background, and what are the channels through which it is most effectively reached with information. For example, decision-makers at the national or regional level may be most effectively reached by television, newspapers or a panel discussion on the issue in a club of which they are members. Householders may be reached through radio, local leaders and extension workers

- ***What type of information is to be communicated***? For example, why-to information on a new idea can be communicated through the radio and a village meeting. How-to information can be communicated through written or illustrated materials, such as booklets and posters, to be explained by development workers, teachers or local leaders. A sequence on how to make something is more suitable for a pamphlet or a flip-chart
- ***Who is going to use the materials?*** Are the materials expected to work by themselves? In other words, is a poster simply going to be placed on the wall of a community hall, or is it meant for discussion with a teacher or an extension worker? And
- ***The budget of the organization or project.*** Making a television programme or a glossy colour poster is expensive. Making simple black and white printed materials is cheap. The decision on how to allocate the budget should be based on a careful assessment of the communication needs, and where it would be most cost-effective to spend the money.

Communication channels can be categorized as:

- ***Mass media.*** Television, radio, newspapers
- ***Audio-visual and print media.*** Posters, booklets, pamphlets, flip-charts, slide-shows, audio-tapes, and so on
- ***Local media.*** Traditional story-tellers, folk singers, puppet theatre, local shows and demonstrations
- ***Interpersonal channels.*** Teachers, trainers, extension workers, builders, apprenticeships, leaders, and
- ***Reality.*** Buildings, building sites, building component workshops.

Another useful distinction relates to the purpose of the materials:

- ***Promotional,*** and
- ***Teaching or instructional.***

Promotional materials are meant to work by themselves, that is without the help of an intermediary. Such materials do not have a built-in process of feedback, and do not involve interaction. Examples of promotional materials are a television or radio programme, a poster in a public place, or a pamphlet distributed to all community members. The information in promotional material usually needs to be quite simple and easy to remember, and so should not be over-burdened with too much content.

Teaching materials are made for use by instructors with a group of the target audience. Such materials are meant to stimulate the process of communication between the instructor and the audience, and usually involve feedback. Examples of teaching materials are: a video programme which introduces an idea for discussion; a radio cassette which an extension worker plays with a group to start a discussion or a demonstration; a poster illustrating an issue which needs further elaboration by a teacher; and a flip-chart illustrating the steps in a process where an instructor provides the explanation.

Mass media have the potential to reach large audiences. However, the information cannot be very audience-specific. Also, there is usually no way to get feedback from the audience, and thus know how the message has been received. The mass media is a one-way information channel which is commonly used for promotion of new ideas. Where decentralized mass media is available, such as community radio

Case study: 'Think before you build' in Yemen

After an earthquake in Yemen in 1982, Oxfam, Redd Barna, and Concern established a small building education project. The focal point was a purpose-built training centre which itself was a demonstration building for the techniques being promoted. A mobile team travelled to villages to raise awareness, with the help of short videos and discussions, and to encourage masons to come to the training centre. Later in the project temporary rural training centres were established which would stay for up to four months in one location before moving on. In the four years of the project over 1000 builders were trained. In the following extract from an article in Open House International *(Vol.12, No.3, 1987) the project director, Jolyon Leslie, considered the efficacy of the various educational materials employed.*

Educational materials to be used for training at the centres needed to illustrate some basic technical points, including:

- The importance of level and stable foundations
- The advantages of using properly mixed cement mortar
- The benefits of good stone bonding
- The need to consider the size and position of openings in walls
- The importance of continuous horizontal reinforcement, and
- The need for strong connections between walls and the roof.

A simple printed booklet was prepared illustrating these points with line drawings. It serves as a kind of textbook for the training, but has also proved useful for the mobile workers to introduce ideas to builders they meet in the villages.

Some of the more abstract images (that is, parts of walling shown in detail) have proved more difficult to understand than drawings that show an entire house. The use of colour (red) to emphasize parts of a detail, such as reinforcing bars, seems to be effective. The device of contrasting wrong with right techniques by using crosses and ticks, is quite widely understood, if the contrast between images is vivid enough. Subsequent drawings have incorporated some of the lessons learned from the use of the booklet. Images are larger and deal with only a single technical point, so that trainers might easily use them to stress some part of the course. Where details have to be used, a familiar object such as a hand or tool is incorporated to give the image scale, and offer a clue to the subject. No drawing is used without an accompanying verbal explanation from the trainers.

Cartoons were used to try and convey a building process which required a number of different steps. The idea of sequence has proved to be a very difficult notion to convey on paper, even when pictures are numbered. Such methods are more effectively dealt with on film.

Posters have been used to give information about the activities of the project, rather than cover technical issues. They show damaged buildings and people at work on new construction, and bear a slogan used in the training course ('think before you build'). Posters have been put up in village schools and local government offices. They have also been fixed to the cabs of trucks that travel regularly to more remote areas.

Black and white photographs, mainly of typical damage to traditional buildings, have proved useful in the training courses. These are generally understood when human figures are included. However, constructional details in close-up are harder to grasp. Colour photographs attract more attention than monochrome, and in some cases seem to be more easily recognized.

Slides are presented in the training courses to illustrate damage and show reinforcing techniques. They are particularly useful in emphasizing a single aspect of a building, for details seem to be more readily understood when projected than on paper. Slides, like other materials, have been continually replaced as new issues emerge from the training.

Video films were initially to form the basis of the training programme. They have been used for village work, and short films about the construction of a demonstration building are shown as part of the training courses. Their value would seem to lie in their use as an awareness raising medium, rather than to communicate technical points about construction. This might have some relation to the familiarity with television, which is seen primarily as a source of entertainment, rather than information. A short humorous film made by the project staff about the short-sighted attitude of those who build poorly is very popular with the trainees.

Simple wooden models have been made to demonstrate the mechanics of building failure, such as the cracking of corners. The models are hinged to fail, and the trainees are encouraged to suggest ways that the elements of the building might be strengthened. Despite the potential problem of scale, with villagers expected to visualize a wooden object as being a building, the models are popular with trainees and seem an effective means of conveying ideas about collapse.

or local newspapers, feedback can more easily be built into the long-term production of programmes. It is worth noting that the educational effect of television is questionable. In Bangladesh, it took more than a year of viewing television before people were interested in seeing educational programmes. They saw television as an entertainment medium.

Audio-visual and print media are used for promotion and teaching purposes. The decision on what media to choose also depends on the skills of the intended users. In the case of a slide-show, the skills of the extension worker who will show it to, and discuss it with, the audience will influence its potential effect. Thus, the development of the medium is only the first step in the dissemination process.

A special note should be made about posters. Posters are to many development workers synonymous with communication. Yet, they have severe limitations. They do not speak and they cannot listen. And, they are often assumed to just work, magically, on their own. Research and experience has shown that in most cases they do not. An appropriate use of a poster is to give a simple informative message, like: 'building information is available in the community hall every Monday'. A poster can also be a reminder of a message from another communication channel.

If a process is to be described, media such as slide-shows, videos or flip-charts can be appropriate. Common to these are that they need a person to discuss the process during or after the display of the visual.

Local media are often the most readily available, the cheapest, and potentially carry credibility with the population. Also, the local media are very familiar with the local culture, frequently use humour, and most often involve feedback from the audience. Use of local media will sometimes facilitate greater involvement of the community, as the people participating in these media live in the community and continue the discussion of the ideas with their neighbours. As an example of the use of local media, in Vietnam, traditional puppet theatre is being used to promote typhoon-resistant construction.

Interpersonal channels are the people who bring the messages directly to the target audiences through personal communication. This is a labour and cost-intensive method. But, it is potentially the most effective in terms of introducing and facilitating change at the individual level.

The interpersonal communication skills and teaching skills of the fieldworkers strongly influence their effectiveness. Training might have to be considered if the skills are not good enough. The credibility of the field-worker as teacher is equally important. A young, unmarried fieldworker who has never built and lived in his or her own house will have less credibility with the rural householder than an older man from the community who has had time to learn from life as well as from books. The choice and the training of the interpersonal communicators for a project is one of the most important decisions to be made.

Apprenticeship is another route sometimes available to young craftspeople for training. Development programmes could consider subsidized apprentice ship with master craftspeople as a valuable means of further spreading and establishing new knowledge.

The school classroom is still a major channel for introducing knowledge. Courses in technical education for teenagers and young adults with well-designed course materials and suitable training for teachers have the potential to communicate more comprehensive and complex messages than any programme directed at the general public.

Demonstration buildings

There is no substitute for the real thing. Seeing with one's own eyes that a building built using new and affordable ideas has survived an earthquake or a hurricane is the most convincing lesson of all.

Demonstration buildings are potentially a powerful way of conveying a new idea. When people can see and touch the proposed new construction there is less chance for ambiguity and misunderstanding about what is being discussed. The best demonstrations are those which can be carried out on existing buildings which are lived in by respected local people within the community.

Where possible proposed new techniques should be used in demonstrations in the way in which they are intended to be used in people's houses. This may seem obvious, but in practice the only feasible way to demonstrate new construction techniques is often in community buildings such as schools, community halls, communal workshops, and so on. Such demonstrations may still be valuable but often they are buildings with dimensions which are larger than those of houses or which are the subject of far greater wear.

An improved floor finish or building block which may be perfectly adequate for a house may not stand up to the battering it receives when used in a school. If the technology is seen to fail in these circumstances then the demonstration may well backfire and put people off using the idea. One of the best building types for demonstrating domestic construction is the health centre, since it tends to have domestic-size rooms and is a relatively controlled environment. It is seen by large numbers of people and is associated with health and safety.

The fundamental problem with all techniques to promote building safety in hazard-prone regions is that shrewd, sceptical people will only be truly convinced when they have seen a so-called improved building survive a disaster. Inevitably, most building for safety programmes occur after disasters. In the case of earthquakes it will usually be decades before there is another serious earthquake. Agricultural programmes can usually demonstrate the benefits of their ideas in one growing season. Building improvement programmes, with or without demonstration, usually require an act of faith on the part of the target audience.

Demonstration buildings may be particularly well suited to cyclone-prone areas. Cyclones occur every year. And severe cyclones hit many places relatively frequently. In these circumstances it is feasible to plan for demonstration buildings which can be seen to live up to the promises of the promoting institution, or not as the case may be. Demonstration buildings have been used in this way in Vietnam (see the Vietnam case study).

A distinction should be drawn between demonstration buildings and prototype houses. A good demonstration building will be used as a vehicle to promote a number of techniques which can be used independently and some of which, if not all, can be used on existing houses. A prototype house is one which the promoting institution is suggesting should be copied as a whole. There are several problems with the prototype house approach:

- ***Too many ideas.*** It mixes up a whole set of technical ideas so that the important messages are buried under secondary or irrelevant detail
- ***All or nothing.*** It can create the impression that one cannot implement one technical innovation

without adopting the whole package. For instance, will a recommended proposed wall construction only work on the type of foundation proposed, or are they two independent ideas

- ***Implicit focus on new-build.*** The prototype house reinforces an implicit focus on the construction of completely new houses as opposed to repairs to existing houses. It is a commonly observed fault with post-disaster programmes that they neglect repair in favour of new-build
- ***Excessive innovation.*** In a prototype house, as with a building manual, designers are liable to wish to be innovative in all aspects of the design. Improved cookstoves, latrines, and improved house layout are common features of post-disaster prototype houses. With each new idea the possibility of rejection is increased. There is a danger that important and useful ideas will be rejected or overlooked due to dislike of a secondary idea, and
- ***Patronizing.*** The prototype house approach is based on the idea that people who have always designed and built their own houses need architects to tell them how to arrange their homes.

Local relevance

The success of your communication strategy depends on the extent to which the audience feels that the message is relevant to them and that the idea or solution is attainable. Most people take advice from someone they respect. Seek out the people who command respect in the community.

Educational materials developed for a global or even national audience are usually too general for people to feel the relevance to their situation. A good communicator can often bridge this gap by applying the idea to the local situation verbally, and by demonstration.

By making the materials relevant to the local environment and the local situation, you have a much better basis for communication and information. This includes taking account of local customs of dress, house design, and tools, as well as local ways of communicating. Within the local context one must consider who are the best people to communicate the message to an audience, either directly or as role models portrayed in a variety of media.

Most people take advice from someone they respect. Different people are sought for advice depending on the type of problem. A householder wanting to improve his or her house may discuss the issue with the local builder, with a leader in the community, or with a neighbour who is slightly better off and who is also considering some changes.

The implication for the development worker is that he or she has to acknowledge that they are not necessarily the person a householder will turn to for advice. If the development worker has spent a long time in the community and demonstrated their usefulness on the subject, then people will likely seek him or her out. Until then the development worker would be wise to find out to whom the community members go for advice on building issues, and seek out the same persons. These persons will usually be open to discussing new ideas – they are often the important channel through which new ideas are introduced and gradually accepted in the community. This process may or may not involve the use of communication materials.

How many messages?

The development programme which successfully introduces one widescale change of practice is doing better than most. The more messages being promoted, the less chances there are of any one of them being accepted.

Change usually occurs gradually, step by step. No new idea exists in isolation. Each idea builds upon what was previously known and helps to establish the context for the next new idea. The idea will only be adopted if it makes sense in terms of what is already known. The process of field investigation should reveal what is already known within each target audience. It should help to suggest which new ideas will have a good chance of being understood and accepted when introduced.

On the basis of this understanding specific messages can be developed. The number of messages should be kept to the absolute minimum necessary to achieve the objective. The more messages that are promoted the less are the chances of any one of them being accepted. People unused to formal learning will retain ideas better if a small amount of information is given at a time. Preferably, there should be no more than one or two key messages at any stage. As people absorb the ideas, they will ask questions and discuss them. In turn, this will lead to opportunities to introduce further secondary ideas. If there are too many messages it may be that the objectives have been set too high and need to be changed.

This rule applies to mass media as well as instructional media. When an idea is presented in, for example, a radio programme, it is better to explore the same idea from several different angles, and then repeat the main issues, than to present a number of new ideas in the same programme.

In many cases a single item of instructional material tries to deal with a large number of ideas. This might be done for economic reasons. It is cheaper to gather all the ideas in one flip-chart and give this to the extension worker than to give him or her messages one by one. To use such a piece of material constructively requires that the fieldworker knows how people learn, and understands that only one or two ideas should be presented at a time. Many field-workers have never been trained in educational methods. Thus, they do not use these materials in the best way.

If programme designers choose to make educational materials which contain a large number of messages, it should be ensured that instructions to the user are printed on the materials and that teaching methods using instructional materials are included in their training curriculum.

Preproduction testing

Educational tools and techniques must be tested and adapted locally to confirm that people understand what is being portrayed, that they understand the message, and, if so, that the message does not offend or alienate.

The preproduction testing or pre-testing of educational materials is a

Case study: 67 messages in Guatemala

In 1976 Guatemala was struck by a severe earthquake. An NGO called World Neighbours had been working in the effected area for 13 years prior to the earthquake. With financial backing from Oxfam and technical support from the consultants Intertect, World Neighbours initiated a community-based reconstruction programme which assisted in the building of several thousands of houses both through material assistance and building education. In the following edited extracts from his final evaluation for World Neighbours in 1982, the project director, Ron Parker, reflected on educational materials, demonstration buildings and the conflict between technical purity and the pragmatic realities of working in the field.

The housing education programme promoted aseismic construction by means of a series of recommendations. These recommendations were repeated and re-emphasized in all phases of programme activities: in the flip-charts and printed materials; in the vocational school classes; during the construction of the model buildings; in classes given by extensionists at the community-level; at the subsidized construction materials centres; and upon occasion, in radio announcements. Villagers who wanted to build their homes as a part of the Kuchuba'l groups were required to follow all of the programme recommendations.

It would not surprise me if no one else who works in the housing education programme would agree with the following statement, but I strongly believe that **there were far too many recommendations!** Not even the extensionists could agree on what they were. Worse yet, I felt that several of the extensionists could not tell which ones were extremely important structurally and which ones were not. This resulted, at times, in the extensionists insisting that home builders put some detail into their new homes that was not really needed.

Almost all the recommendations that the programme promoted contribute measurably to safer housing. They can, each and every one, save lives and that is why they exist in the first place. The problem that I see in them is that having so many recommendations lessens the pro-

On educational materials...

The World Neighbours produced educational materials which were generally conceded to be of very high quality. Their strength lay in the participation of field staff in their planning and preparation and in their having been so thoroughly field tested. World Neighbours realizes, although many other groups do not, that educational materials without an educational programme are almost worthless. Our materials were purchased by more than 100 other groups and agencies. This was fortunate because it kept the local people from being confused by contradictory recommendations. I can foresee a future post-disaster situation in which educational programmes work at cross-purposes. The importance of co-ordinating approaches cannot be overestimated. And good educational materials can be as useful for reaching and influencing other agencies as for aiding instruction in the field.

On demonstration buildings ...

One of our earliest efforts was the construction of almost 60 model dwellings. The existence of model antiseismic buildings so soon after the quake and in so many communities kept people's minds open to the ongoing educational effort. The presence of these buildings made it harder for anyone to rebuild in the old style. People found that they could not justify placing their families in danger when an alternative was so visibly present. Also, the fact that the buildings were there for people to study at their leisure led many villagers to conclude that they were well within the range of their own skills as home builders and their reach economically. In conclusion, the model buildings project was a low-cost, highly visible part of our programme, and contributed greatly to the eventual success of the overall housing education effort.

gramme's impact on the target population. There are too many recommendations to be taught effectively to almost anyone. The difficulty lies in the nature of the task: housing cannot be taught like agriculture, which is a cyclical function with techniques that are repeated and improved upon each year. Reconstruction was essentially a one-shot venture for each family, in which a substantial portion of the family's holdings must be permanently invested. Housing education must therefore be approached as an attempt to change people's concept of what a house is, or should be, before they rebuild.

If it is indeed an error to focus on the recommendations approach, as I think it is, there are a great many people who should hear about it. The thinkers I met at the housing conference (the engineers, architects, mathematicians, physicists, and the like) are persisting with their search for even more recommendations in the pursuit for safer housing. Our experience would seem to indicate very strongly that more recommendations mean less safe earthen housing for the people that need and use it.

The paradox of promoting the recommendations was that the harder the extensionists worked to see that a few people followed all the rules, the more they were forced into an adversary relationship with their neighbours. This prevented them from reaching the number of families that needed safe housing. Almost all of the 67 recommendations are important. Even if you could remove, say 27 of them, and reduce the list down to 40 you would still have more than you can teach effectively. And I think there are not 27 rules that anyone working in housing on a theoretical level would want to remove. Most scientific work being done worldwide is designed to make the list grow! The message from the field is this, plain and simple: one million homeless represents around 200 000 families, and an equal number of homes to be built. It is a pedagogical and pragmatic impossibility to teach 200 000 people 67 things. Period.

process designed to ensure that proposed new materials will be understood and useful before they are reproduced and disseminated in large quantities. It means going to the field, interviewing members of the target audiences, improving the material, and testing the revised version. The testing is repeated until enough people interpret the material in an acceptable way.

Potentially, pretesting can be carried out with any kind of educational media. Videos, street theatre, and school courses can all be evaluated and improved or discarded before their widescale application. But the greatest experience in, and commonest use of, pretesting is with graphics. The objectives of pretesting are to find out:

- ***Literal content***. Do the respondents interpret visual material the way it was intended by the artist?
- ***Comprehension of the message***. Do the respondents comprehend the abstract message the images are intended to convey?
- ***Acceptability***. Is there anything in the picture which offends or alienates the respondent?

The most common use of pretesting is to establish to what extent the contents of the picture are recognized – object by object, and as a complete picture – and to identify which parts need to be changed. Comprehension of a message from a picture alone can not usually be expected from audiences unused to reading pictures.

Pretesting can not assess whether a message will be accepted or rejected by the target group. People may understand a message perfectly well, but not find it relevant or important to them. The method can help to identify if there are elements in the picture which offend people, or which they do not find relevant to their situation. If serious problems are identified, these may be signs that the message itself is wrong.

(a) Pretesting techniques

Pretesting can be conducted both with individuals and with groups. Pretesting with individuals puts the picture to the most difficult test, and is most likely to reveal problems of comprehension. Testing with groups is more likely to result in identification of problems relating to the perceived relevance of the message to that community. Thus, both techniques should be used.

All of the techniques described in the previous chapter concerning interviewing in the field also apply to pretesting. In particular:

- ***Two people***. Pretesting should ideally be conducted by a pair of interviewers, one asking the questions and the other recording the answers.
- ***Explain the purpose***. Before starting the interview the interviewer should take time to introduce himself or herself, their task and their purpose.
- ***Test the pictures, not the people***. The interviewer should stress that the purpose is not to test the intelligence of the respondent, but rather the quality of the picture. This is usually a difficult task for inexperienced pretesters. The tendency to judge people based on their success in interpreting the picture is common. This tendency often prevents the pretester from finding out what the real problems are with the picture.
- ***Open-ended questions***. As a rule, one should ask open-ended questions. Towards the end of an interview, if certain elements of a drawing have not been identified, it can be useful to ask some leading questions. For example, if a drawing shows a house damaged by disaster, but this has not been commented on, one might ask is there anything wrong with the house? This form of leading question can help to overcome a hurdle and

so lead on to another field of enquiry where open-ended questions can again be used to explore an issue which would not otherwise have been reached.

- *Take time.* It cannot be stressed enough that one should take time to listen and learn. A few good interviews are more valuable than a large quantity of invalid data.

(b) How large a sample?

Pretesting is not a statistical exercise, it is a qualitative research technique with a well-defined purpose. It is a tool in the design process. The purpose is to identify the problems in the picture, and to go on testing until one is sure of the nature of the problem and what the next version of the picture should look like. The pretester must in each case decide when he or she has reached that goal. Typically, this could mean anything from 10 to 30 interviews. In some cases, even a few interviews will reveal a major problem that needs to be solved before further testing can be done. If the first five people all experience the same problem it may be time to go straight back to the drawing board. If the target audience involves a broad cross-section of people, it may be necessary to test with a larger number of people to ensure acceptable comprehension.

(c) Who should pretest?

Fieldworkers usually make good pretesters, after having been trained in the techniques. Designers or artists will often find it difficult to be unbiased: it is hard to accept criticism, especially from people with no schooling. However, artists and designers can benefit greatly from participating in pretesting – as observers or recorders, without permission to speak or to reveal that they have made the drawings. Listening to how people interpret their pictures, or fail to understand their intentions, is a very good learning exercise for the artist.

A conscientious designer, once convinced of the value of testing, can find the testing process stimulating and challenging. Furthermore, direct contact with the audience may suggest alternative ways of presentation, which can be sketched and tried out immediately. The important thing is to regard the drawing as a tool rather than as a finished piece of art. It is also important not to reveal the identity of the artist until confidence has been built up with the respondents and they have identified the problems with some of the drawings. Once this is done, and the respondents see that the artist is happy to make new sketches based on their suggestions, the co-operation is usually very positive and dynamic.

Similarly, it can be a life-changing experience for planners and policy-makers to participate in pretesting as observers. Through coming to under stand the problems of communication at the level of individual householders, they can begin to appreciate that communication is a major problem requiring time, resources, staff, and training.

(d) Limitations of pretesting

Pretesting of prototypes or drafts does not take into account:

- *Context.* When a drawing is seen in an interview, it is seen out of context. If someone can decode a picture in an interview, it does not necessarily mean he or she will decode similar pictures when encountered casually in the street. Similarly, failure to understand a complex image of a construction sequence does not mean that the same image could not be understood when a teacher is leading an audience through the drawing

Figure 1. A diagram used to explain tuberculosis in Nepal. When used on its own it was unintelligible. Yet, when used as a teaching tool its meaning was understood and remembered.

- ***Learning***. Because an audience is found not to understand a drawing convention, this does not necessarily mean it should be rejected. It could mean that a concerted effort should be made to teach people the convention. The tick/cross convention is one such example. Even quite complex and abstract graphics can be learned in the right context. The drawing in Figure 1, which is designed to explain the transmission of tuberculosis, was tested in five villages in Nepal and understood by nobody. In a sixth village large numbers of people understood it. It emerged that the diagram had been used by a visiting health team several months earlier.

- ***Finished products***. Pretesting assesses approximations to the finished product. Small details can confuse when care is not taken. Details to which people respond are sometimes unexpected, and may not be revealed by the testing process. In the context of an earthquake sequence, people were asked which of the two houses in Figure 2 might be stronger, to see if they would notice and speculate on the coloured ring-beam (coloured red on the top house). Nobody noticed the ring-beam, but two respondents said the lower house was stronger since the central upright of the window frame was thicker. Pictures to be pretested should be drawn as close as possible to the finished product, in content as well as in style.

Field-based preproduction testing of pictures has contributed to opening the eyes of planners, policy-makers and designers that the skill of reading pictures is dependent on culture, background, environment and experience. It has also shown that people usually have a very good reason for interpreting pictures the way they do. Pretesting is a method that opens up the possibility of understanding the perceptions and ideas of others, and thus entering into a constructive dialogue with them – with or without pictures.

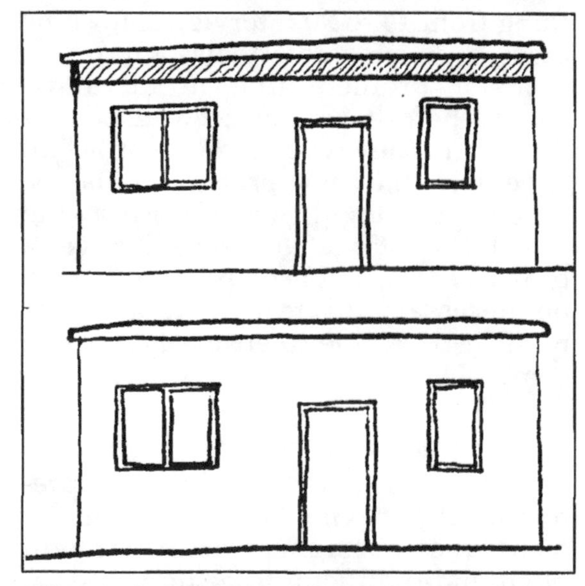

Figure 2. Which might be stronger? The artist's intention was to portray the upper house as having a ring-beam and thus as being stronger. In practice, nobody noticed the coloured ring-beam, while two people commented that the thicker central upright of the window frame in the lower picture indicated a stronger house.

Educational context

The designer of educational materials needs to consider the educational context in which the materials will be used. A technique which is suited to one context may be inappropriate in another.

In building improvement programmes different educational materials are suited to different educational situations, such as:

- ***A flip-chart*** being used by extension workers with small groups of householders (five to 15 people). This implies two-way communication with discussion. The flip-chart could be in two parts – why-to information in the first part, with a clear instruction to the user to stop after this part, and how-to information in part two, which would be the subject for a new discussion.
- ***A slide-show*** being shown to a large audience (50 to 100 people), implying largely one-way information with few questions and comments. This would be suitable for giving why-to information.
- ***A brochure*** to be distributed to the community, designed to be read aloud by literate members of the community to small groups of people, to discuss ideas. The contents could be similarly divided into why-to and how-to information, with suggestions to the reader to divide the discussion into two separate occasions.

Materials designed for more formal education, such as training programmes, can be more complicated in their design and content. Here, trainers can use the materials with trainees over time, and have opportunities to discuss, re-explain, and practise what has been learned, and assess what the trainees have understood. However, here also there are strong reasons to develop clear, simple materials, at least for parts of the course:

- ***Continuing dissemination***. If trainees are supposed to go back to their communities and implement their new skills, it will assist them greatly to have good visual materials to support their explanations. The materials may also help to keep people's attention and potentially to achieve greater credibility with the community members. Using the materials with which they have been taught themselves may help them transfer the ideas better.
- ***Limited skills of the trainers***. Many trainers are not very skilled at using complicated instructional materials. Teaching with visual aids is not a subject which is commonly taught in vocational schools.

Thus, the planner and designer of the educational materials need to look carefully into the educational context before making decisions about the content and form of the materials.

Training

Graphical material rarely works effectively on its own. However good the educational materials, the most important thing is the interpersonal communication skills of the fieldworker using the materials.

It cannot be stressed too often that educational materials are simply an aid to teaching. The principal factor determining success or failure is the skill of the fieldworkers. Projects concerned with developing educational materials for use by fieldworkers need to assess the skills and attitudes of these workers carefully before developing the materials. If skills are lacking, the project has two choices. First, facilitate training of the fieldworkers. Second, choose a different channel for communicating the ideas to their audience.

The belief that a visual aid in itself will do the job is seductive. Unfortunately, it is not true. It usually needs people to help communicate the ideas. While a video film shown to a village might be greatly appreciated as a novel entertainment, unless there is follow-up conducted by people who are trained and motivated, the effect will be very small. The follow-up may be in the form of discussion, distribution of materials reinforcing the ideas, repetition of the video after a couple of weeks, or some similar technique.

Lack of skills in communicating ideas, with or without the use of visuals aids, has been identified as a major problem in the field. The skills are lacking in communicating to different levels of people: to decision-makers; fieldworkers; community members; or others. Dealing with this problem is frequently the largest challenge for development programmes. And it is a problem which technology cannot solve.

Impact assessment

However good the preproduction testing of the educational materials, there is a continuing need to monitor the impact of the materials in use.

Finding out which educational inputs contributed to a certain change is very difficult. The main reason is that a person is influenced from so many different sources, often without being conscious of it. To identify exactly what caused a behaviour change is often impossible or, at the least, requires long-term research which is beyond the scope of most development projects.

However, there are methods that can be used to measure the likely effect of educational interventions. One method is focus-group discussions with key members of a community before an intervention is used, and at various intervals during and after implementation. This method will provide in-depth qualitative information. It has the additional positive effect that it places the respected community members in charge of the feedback to the development agency. There is a possible disadvantage to this method. This is that if the community leaders have an interest in giving or concealing certain kinds of information for their own reasons, they can do so. This will usually be the exception rather than the rule, if the relationship with the development project is good. It should not be used as a reason against using this method.

A building improvement programme also needs regular feedback about responses to new technological options from the community. Such monitoring mechanisms need to be built into the project design from the start. This also demands that the project actually has the flexibility for making adjustments in policies and activities. Channels for relaying such feedback should represent several different economic strata in the community.

Summary: **Educational materials**

- ***Concentrate on one or two essential messages.***

  The development project which successfully introduces one widescale change of practice is doing better than most. The more messages that are promoted the less are the chances of any one of them being accepted

- ***Adapt educational techniques locally.***

  Educational tools and techniques must be tested and adapted locally. Although there are no universal solutions to communication tasks there are universal problems

- ***Identify clear targets and educational contexts.***

  No single educational tool, technique, or channel is going to be adequate for all audiences and tasks. All educational material should be designed with a particular audience and context in mind

- ***Use the real thing.***

  Demonstration buildings can be the most effective way of communicating improved construction

- ***Invest in staff.***

  Staff at all levels need to be aware of the importance of communication skills. Investment in training of field staff in developing, testing, and using educational materials is vital.

Case study: **Multimedia in Vietnam**

In Vietnam, the United Nations Centre for Human Settlements (UNCHS), with funding from the United Nations Development Programme (UNDP), recruited Development Workshop/GRET, a consortium of non-profit organizations, to assist the Government to promote typhoon-resistant construction. Working with the community in a series of workshops the project identified 10 key messages and developed a wide variety of media. Any one of the communication techniques would in itself have been inadequate, but, taken together they formed a powerful and mutually reinforcing campaign. In their final report John Norton and Guillaume Chantry, of Development Workshop and GRET respectively, summarized their approach.

The workshop participants have been guided to examine locally applicable ways to communicate information about typhoon-resistant construction techniques. On this basis they have developed communication materials aimed at the general public, at builders and technicians, and at decision-makers in the different provinces concerned. These materials comprise posters, videos for explaining the programme, games, radio announcement texts, and a film *Our House Resists the Storm* filmed in Phu Loc as part of the project, and used to advertise the role of the district technical adviser and the 10 key points of typhoon-resistant construction.

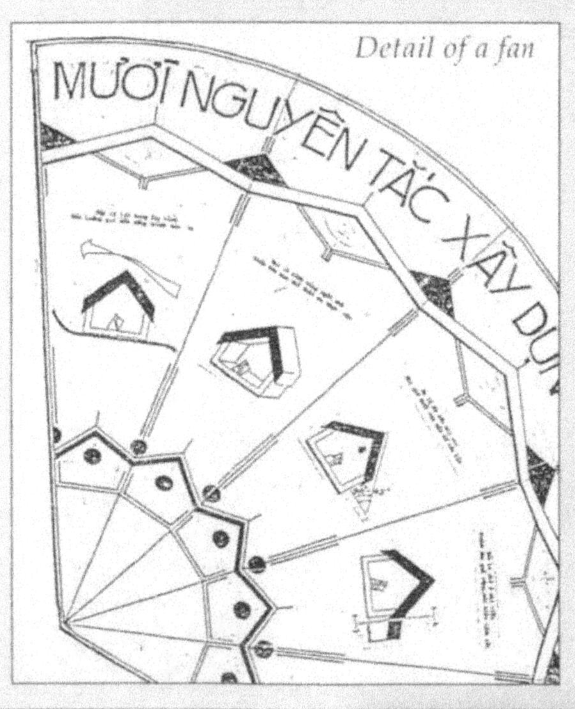

Detail of a fan

These materials have then been used in two public education programmes, the first as a trial in Phu Loc District, Thua Thien-Hue province, and the second at provincial level throughout Thua Thien-Hue. They have been dynamic experiences, which have raised the awareness of the general public, and of decision-makers and technicians among government departments and institutions.

In Phu Loc the campaign mobilized local institutions including the People's Committee, the Technical Services, the Red Cross brigades in the schools, and the Disaster Preparedness Committee, and used local services including district mobile videos and the local radio service. The media developed in the seminars were tested. The results were encouraging. In two weeks more than 5000 people saw the video, numerous posters were distributed and commented on. And the programme participants gained ideas about how to improve the campaign.

In the case of the provincial education programme, activities have been multiple and rich. In the Hue, the newspapers ran full page articles and press releases prepared by the Government's core team. The core

Playing cards for safety

team toured the province to lecture and discuss, and appeared on local television with provincial leaders to carry the message further. Provincial and district radio announced the programme and the time of showings for the video (*Our House Resists the Storm*), watched each time by several hundred people before the showing of main feature films, which are already a regular event. In localized activities, photograph and drawing exhibitions showing activities against natural calamities were organized. In the schools competitions were held for poetry and drawing about the Campaign for Typhoon-Resistant Building. And throughout the province the Women's Union, The Youth Union and the Farmer's Union organized public gatherings on the same theme. All over the province more than 2500 large posters were shown in the main places of gathering, the markets, bus stations and cafes. On the Provincial Day of Disaster Preparedness, the 26th April, the youth brigades paraded in the streets of each district with specially prepared banners, and the radio and TV ran special programmes. This campaign met with great enthusiasm and several districts prolonged activities into a second month.

Summary of materials

- Full colour posters illustrating the 10 key points of typhoon-resistant construction
- The video *Our House Resists the Storm*
- Loudspeaker and radio announcement texts – on typhoon resistance, and on times of showing the video. Megaphone announcements for publicizing the video, used in the districts
- Television presentations
- A folded leaflet showing the 10 key points of typhoon-resistant construction
- A manual for technicians showing the 10 key points of typhoon-resistant building
- Technical dossiers for each of the three provinces on local techniques to be encouraged to achieve typhoon-resistant buildings
- A dice game based on the traditional horse racing game played with a board illustrated with the 10 key points
- A pack of cards illustrating the key points
- The design of a fan showing the 10 key points
- Five poems about typhoon resistant design
- Three demonstration buildings, and
- An exhibition mounted during the public information campaign in April 1990 by the Government's core team.

3. Illustrating building for safety

Printed images for building education need to be designed on the basis of an understanding of how people from the target audience interpret pictures. Although the problems of visual interpretation will vary from place to place there are some important problems which commonly reoccur.

There is a large range of educational materials which are suitable for building improvement programmes, depending on the background of the target audience and the nature of the messages. However, there are several reasons for focusing on printed materials:

- They are very commonly used
- They are widely distributed within projects
- They are often seen by people in other projects and countries
- They are readily reproduced using simple, widely available equipment, and
- They are readily pretested with the target audiences.

Many projects concentrate on developing a variety of printed materials to support their activities, from posters and leaflets to flip-charts and manuals. The previous chapters have discussed how to choose the appropriate material for the task. Assuming that the project has concluded that printed materials are called for, it is time to analyze in detail what should guide the development of the illustrations of such materials.

People who have gone to school and live in an urban environment are usually highly skilled and practised in the art of interpreting pictures on a page. The skills which someone uses in interpreting drawings in books, photographs in newspapers, and images on advertising hoardings all have to be learnt. Once learnt we are rarely aware that we have these skills.

Many people in developing countries, especially people who live in isolated rural areas, have had very limited exposure to pictorial material. The ability to interpret pictures is sometimes referred to as visual literacy. Those unable to interpret illustrations are described as visually illiterate. This description reflects our own bias. These people may be perfectly able to interpret images, such as religious ones, which are familiar to their own culture and surroundings but whose meaning is obscure to outsiders. They have a degree of visual literacy regarding these images which a person from another culture would not have. However, they are not familiar with the kinds of images and symbols which are taught in schools and are part of a visual tradition that many development workers take for granted. Given the right circumstances, people can learn to interpret certain types of graphical conventions quite quickly.

Research has been conducted in several countries on how people of limited literacy interpret graphical material. The principles derived from this research have been tested out on building improvement topics by the Building for Safety project. The following sections summarize some of the issues which emerge. These can help to guide the process of the local development of graphic material.

Picture styles

People unused to reading pictures tend to interpret marks on a page very literally. The representation of an object should provide as many clues as possible to the object's identity. Spurious detail which does not provide clues should be omitted.

People who are unused to reading pictures will interpret the images very literally. This means they will usually:

- Look for the concrete meaning rather than an abstract one
- Look at each part of the drawing separately, without necessarily assuming a connection between the different parts, and
- Look at the picture from their own local viewpoint trying to make sense of it for, or from, their own environment.

(a) Abstract ideas

If images are interpreted literally, how can you communicate abstract concepts and ideas to people in graphical form? The answer is you do not. Illustrations of abstract ideas usually do not make sense to people unused to reading pictures. This result has been found in several studies on visual perception. The following examples from northern Pakistan convey the problem. The task was to illustrate the concept of strength and more specifically that building a ring-beam on your house would strengthen it or hold it together.

In Figure 3, a belt is used to suggest the idea of holding the house together. Most of the people had no trouble recognizing the house and the belt. But, beyond that the most common response was 'we do not build like that here'. The drawing was being interpreted literally, and the idea that the picture might have a meaning seemed alien.

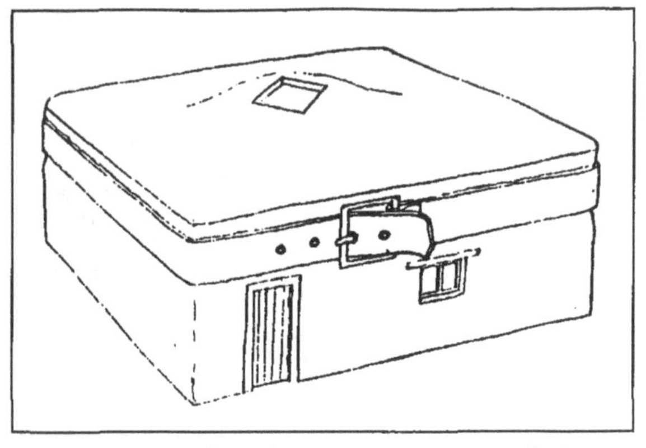

Figure 3. A failed attempt to use a belt as an analogy for a ring-beam.

In Figure 4, a muscular man is used to suggest the idea of holding the house together. Few people had problems identifying the literal objects of a house and a man. Several people observed that the man was strong. However, the conflict of scales within the image was so evidently nonsensical that many people were forced to think again: something was not right with the drawing. A few people revised their literal interpretation and

Figure 4. Strength conveyed as a strong man holding the house together.

Figure 5. Casting a ring-beam on a stone house.

suggested it was a model house or a box. Some people suggested the drawing showed that the man was proud of or loved his house.

Neither of the drawings succeeded in communicating the idea or the intended message to the audience. However, the scene of the man holding the house was so obviously unreal that it seemed to be more successful in provoking people to search for another interpretation. The house with the belt was, though odd, not completely beyond the bounds of possibility.

Successful communication of abstract ideas or concepts through illustrations usually needs a person to explain the ideas. And, even then, one cannot be sure the idea is understood. The illustration might confuse rather than illuminate as such drawings are alien. In a case like the one above, of trying to explain how a ring-beam makes a house strong, it would be better to:

- Use a simple realistic drawing of a ring-beam, such as in Figure 5, and
- Show the real thing or a similar idea, on a house in the neighbourhood, and discuss the pros and the cons in the local setting.

The desire to interpret images literally is generally the overriding principle for people of limited literacy and should be constantly kept in mind by designers of educational materials.

Figure 6. Perspective drawing used in a Unicef study in Nepal.

(b) Perspective and three dimensions

In building education materials, the use of three dimensions and perspective is common and necessary for illustrating a number of concepts and ideas related to building houses.

Reading perspective and other representations of three dimensions is initially problematic for people with low skills in reading drawings. However, research in Nepal has shown that people can quickly learn to interpret perspective. A study by Unicef assessed how rapidly people could acquire new skills of pictorial interpretation. Of the 10 different categories tested, the understanding of perspective and background, as used in Figure 6, went up from twenty-four to ninety-two per cent over the course of six visits. The study does not assess how other criteria to facilitate such learning, such as the attitudes and communication skills of the interviewers, influenced the results.

In a study with 270 respondents in Lesotho only twelve per cent saw the top drawing in Figure 7 as three-dimensional. In another drawing where the steer's head slightly overlapped the house, only eight per cent comprehended the three-dimensions. In the lower drawing, thirty-eight per cent of the respondents read it as three dimensional. Almost everyone could identify the house, the steer and the donkey in the two pictures, but only twenty-one per cent recognized the mountains.

The researchers concluded that the potential for a three-dimensional drawing on its own to communicate a message successfully was very small. They also concluded that giving clues on the drawing, by adding a road and details on the ground, significantly improved the ability to comprehend the three dimensions in the drawing. However, due to photography more and more people are becoming used to the conventions of perspective and even where they are not the Unicef study suggests that people can rapidly learn.

Some examples from building education materials illustrate the lack of awareness among designers about the difficulties householders may have to interpret three dimensions.

Figure 7. Perception of depth in Lesotho. While thirty-eight per cent of the respondents read the lower drawing as three-dimensional only twelve per cent did the same for the top picture.

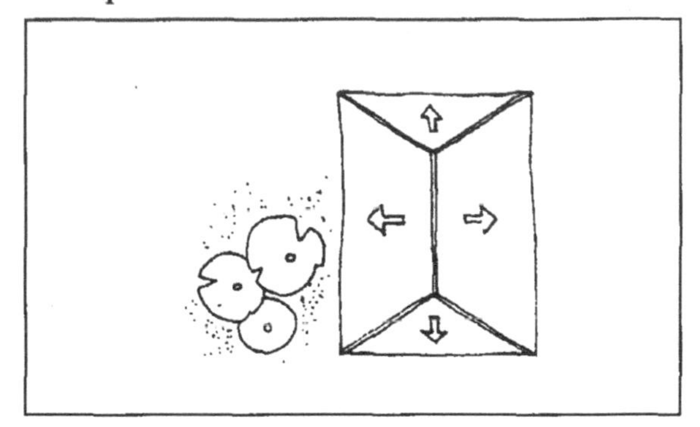

Figure 8. A drawing of a house from a booklet produced by a UNCHS project for rural communities in Ecuador. The unusual viewpoint, the oversimplified representation, the use of symbols and irrelevant detail, all conspire to cause confusion.

Figure 8 was produced to describe a roof which slopes in four directions. The possibility is high that rural people did not comprehend this image. Furthermore, the trees to the left of the house have no purpose in the drawing, and their presence, as well as the unusual viewpoint, would contribute to the confusion.

Figure 9 will most probably be meaningless to most rural householders and rural builders. In addition to using perspective, it uses the speech bubble of the cartoon, implying the

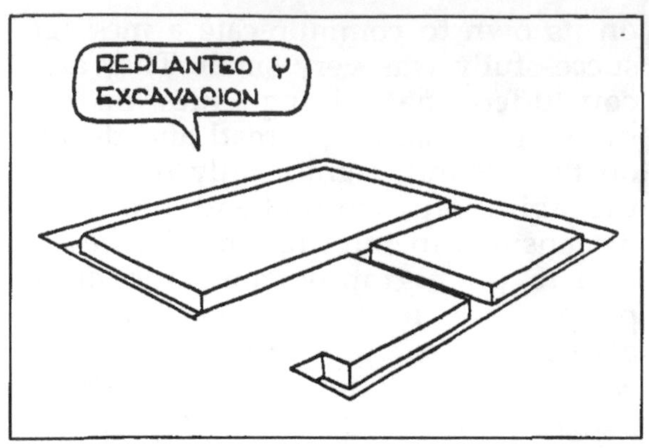

Figure 9. Drawing from a promotional booklet depicting the foundation trenches for a house. The oversimplified representation and the use of a speech bubble will probably cause confusion.

abstract and completely nonsensical idea that this structure has something to say to you.

Figures 10 and 11 are both taken from construction manuals produced for post-earthquake reconstruction projects in Ecuador. Of the two, Figure 10 is much easier to read. A comparison between the two implies the following principles:

- ***Shadows.*** The simple but careful use of light and shade can give a drawing a clear three-dimensional form. Too many or too strong shadows can confuse
- ***Scale.*** Recognizable objects, in particular people, can be used to indicate scale. Once the viewer has recognized the clue, he or she will more easily realize that they are looking at a house-sized object, and
- ***Viewpoint.*** The angle of view should allow the reader to see the necessary structure and details, with the minimum amount of distortion. Figure 10 is a good example of this. A higher viewpoint could present the picture like a plan, but not quite being a plan, it could be confusing. A very low viewpoint, as in Figure 11, can make objects such as walls in the foreground clash with objects further back.

The use of three dimensions and perspective is often necessary in building education materials. The designer should take care to make the drawing as realistic as possible, providing clues that help interpretation, and avoid abstractions.

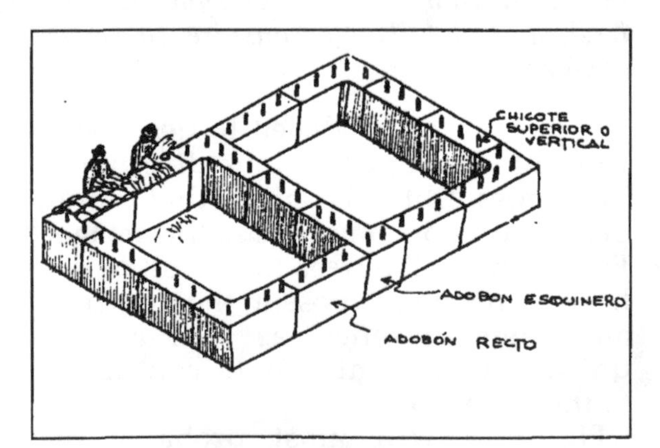

Figure 10. The shadows, the people, and the careful choice of viewpoint help to make this isometric drawing of a house under construction easier to understand.

Figure 11. The line drawing technique and the choice of viewpoint may lead to confusion in interpreting this perspective image of a rammed earth house.

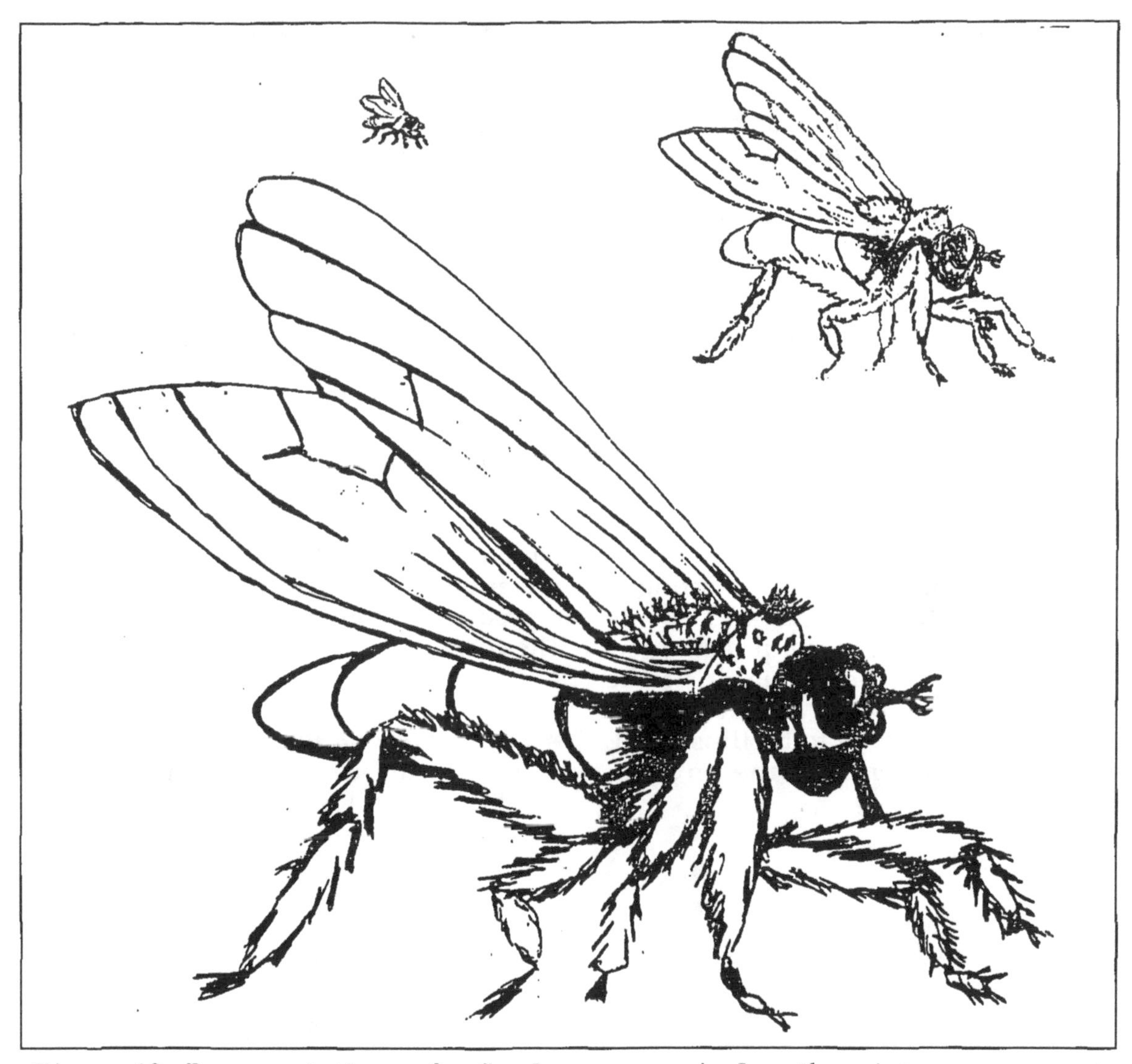

Figure 12. Representations of a fly. In a survey in Lesotho, sixty-one per cent understood the small drawing, forty-seven per cent the medium sized one, and twenty-seven per cent the large one.

(c) Scale

Objects represented at a scale very different from reality can sometimes pose a problem to the viewer. A classic example is that of a health worker in Africa who showed a big picture of a fly, to explain how flies are a threat to health. The people walked out after the presentation with smiles of relief, saying 'We don't have a problem, because our flies are just very small'.

Sometimes it is necessary in educational materials to enlarge a detail to make it more clear. In such cases you should:

- Be aware that the drawing by itself may not be understood by illiterate people
- Explain to the user of the material that research has shown this drawing to be complicated, and be sure to explain that the reason it has

been drawn to a larger than natural scale is to make it more clear, and

- Where appropriate, make a drawing of the whole situation or context to which the detail relates and show the link to the detail with an arrow. But note that this method is also not easily understandable by itself. It will usually need somebody to explain the connection. Advice on this difficulty should be given to the user in the text.

(d) Graphic technique

The main picture styles used for images on paper in educational materials in the developing world are black and white photographs and line drawings.

If local reproduction and printing is of good quality, black and white photographs can be a useful technique. Photographs can be valuable if the whole context of a situation is important to the issue to be illustrated, such as the influence of the environment on a village.

Often in building education it is necessary to show a detail or an operation in the context of the building as a whole. For some purposes, deleting unnecessary background by cutting it out will improve the comprehension of the picture. Cut-out photographs can be particularly useful for representing people since they are so difficult to draw well. Sometimes accuracy in depicting people is necessary in order to indicate someone of a particular ethnic group, class, or profession, or for showing someone with a certain personality or mood which is important for the theme.

However, there are, several issues which influence the successful interpretation of photographs, such as those reproduced in Figure 13 and Figure 14. These are:

- ***The camera angle.*** The angle can contribute to distortion of the subject, as in Figure 14.

Figure 13. Photograph of a well-built house of stone and cement. A few people commented that this house was well built. On occasions the window caused confusion, as did the building materials in the foreground. The strong shades and the cluttered foreground and background in this photograph make it unsuitable for educational materials. Reproduction of pictures like this usually makes the problems worse because the shades and the contrast will appear even stronger.

Figure 14. Cut-out version of the photograph in Figure 13. The removal of the background emphasizes the strange angle of this photograph, and made it look like a coffin to one respondent. A couple of people did not recognize this as a house. One described it as a plan of a house compound. The window and the piles of material in the foreground caused confusion.

- ***The light conditions.*** Strong or deep shadows can cause confusion, as in Figure 13.
- ***Irrelevant detail.*** Objects in the picture which are not connected with its main theme can be distracting. In Figure 13 and Figure 14 the children, the heap of earth, the tree, and the mountains were all sources of confusion. The children in the foreground were crying and this was sometimes taken as indicating there was something wrong with the house.

In the few cases where colour photographs have been tested with people not used to interpreting pictures, they have consistently been found to be significantly easier to understand than black and white photographs or line drawings. The technology for good quality colour reproduction is rapidly becoming more widely available and less expensive. It is, however, still considerably more expensive to reproduce colour photographs than black and white ones, and very much more expensive than printing line drawings.

A detailed line drawing is often the best and most economic choice. It is easy and cheap to reproduce, does not need sophisticated equipment for printing, and can be photocopied. But its advantages are not just economic. A good draughtsperson can pick out and emphasize important details which might be lost in a photograph (see Figure 16 overleaf). The more detailed and realistic the drawing is, the better the comprehension, so long as the detail is relevant. However, even simple line drawings will often be comprehended, if the drawing is realistic and contains enough detail to facilitate understanding. When drawing buildings, building details, and tools, the accurate line drawing is usually the most appropriate. To achieve the best possible results, the drawings should be done in black ink.

It requires certain skills to make line drawings. One much-used technique by artists or designers with low skills is to take a photograph, enlarge it, and copy the shapes, shadows and details one needs. However, care has to be taken to copy only detail which helps to clarify the image. Otherwise, problems can arise as in Figure 15.

Figure 15. In this drawing the content of the photograph in Figure 13 was reproduced more or less exactly with little selection of what was or was not significant. The stones were not drawn clearly but in a manner indicating a general texture. Some people commented that the house was poorly built, most probably because of the appearance of the stonework. While in the photograph, the stonework is no more than a texture of light and shade, it still gives people the impression of a well-built house.
In a drawing, a line is expected to be there for a purpose. Stonework represented by broken or incomplete lines was read as poor stonework. Similarly, both the heap of sand and the crying children, because they were carefully drawn in, took on an even greater significance than in the photographs. Respondents commented: 'if someone has gone to all the trouble of drawing it then it must mean something'.

Figure 16. In a survey in Nepal, twenty-eight per cent of the respondents understood the photograph. Ninety-four per cent understood the detailed or half-tone drawing.

Sometimes, the absence of an artist with reasonable skills can force a project to resort to stylized drawings such as Figure 53. If the target audience unused to pictures and the drawings are too abstract or crude they may result in more confusion than illumination. If the materials are to be printed and distributed, and even partly expected to be comprehensible by themselves, it is worth investing time and money to make good drawings. However, even stylized ones are often better than none, especially when there is someone present to describe and discuss them.

For instance, when a non-artist is in a situation where he or she has to explain something to a group of householders, making crude drawings on the spot can often be useful. Also, in discussions, people can sometimes be encouraged to develop drawings themselves to help formulate and express their own thoughts.

Research with audiences of limited literacy in a number of countries has shown that the detailed line drawing and the cut-out photograph are the most easily comprehended styles. The simple line drawing, that is one which is realistic, but without much detail, is the third best understood style. This is followed in fourth place by the photograph complete with background. Fifth is the stylized drawing.

Figure 17. Casting a reinforced concrete ring-beam on a stone masonry house. This is an example of a subject suitable for illustration with detailed line drawing.

(e) How much detail?

Because people interpret pictures literally, a picture should include necessary details which provide clues to its interpretation. It should exclude unnecessary details which can confuse the observer.

Making educational materials for audiences of limited literacy does not necessarily mean making the drawings overly simple, that is to say, making broad outlines of subjects and excluding details. Although this kind of drawing is often preferred by development workers, because it does not require much artistic skill, it can be confusing.

Figure 18 is an example of a drawing that caused confusion because of lack of necessary details. The artist intended the drawing to convey construction of a concrete ring-beam on a local house. When tested with villagers in northern Pakistan, some people described the house as having three rooms. Architects and artists are used to the convention of a few well placed lines describing three-dimensional form. In this case, the lines confused, and were interpreted in the most literal manner available – as dividing walls.

Some people considered the house to be strong because the smooth white walls were believed to signify concrete walls. In northern Pakistan this has social and economic implications; in rural areas, less well-off people build with stone and the richer ones use concrete. In this particular location, drawing the stones on the wall of this picture would make it comprehensible as a local house for common people.

Figure 5 shows a drawing where the stones are drawn in quite carefully. The artist's intention here was construction of a concrete ring-beam on a local house built of stone, with larger stones used at the corners to strengthen the house. This drawing conveyed the

Figure 18. Outline drawing depicting the construction of a ring-beam in Pakistan. It was sometimes interpreted as a three-roomed house. The plain white walls were read as being made of concrete and therefore strong.

intended idea quite successfully to people in northern Pakistan. Not only did they interpret the stonework, many also picked out the larger stones on the corners as indicative of a well-built house. The extra detail made the main message more comprehensible and allowed scope for secondary messages.

Problems are also encountered when artists add extraneous detail which is meant to add interest or to be humorous. Some artists have added animals to drawings to reflect the character of a rural community. But people looking at the drawings have often drawn false conclusions, such as that cow dung should be used in construction or the buildings depicted are animal houses. Figure 19, taken from a post-earthquake reconstruction manual, is intended to show a cracked wall. The drawing is difficult enough

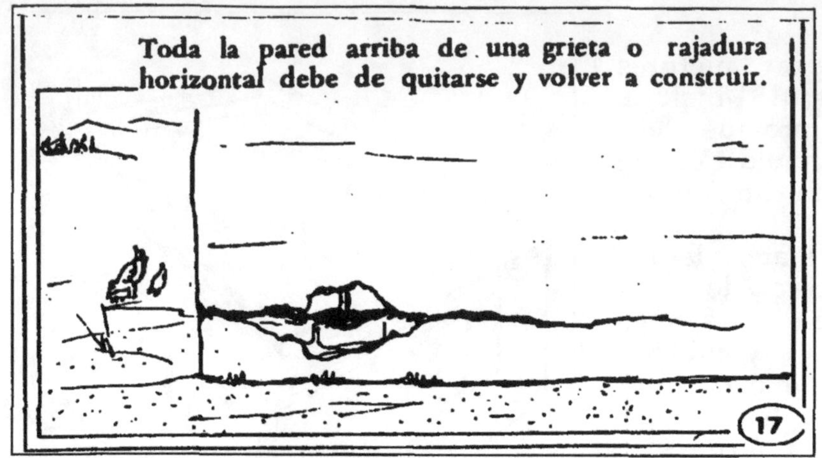

Figure 19. Why are there chickens?

to read already without the addition of two chickens. By placing the chickens in the background they do not even help to establish scale.

This type of unnecessary detail, which is not part of the main theme of the drawing, tends to confuse people who are unused to interpreting pictures. Adding the stones on a house is helping the comprehension, as the stones are a natural part of the house and give another clue to the theme. The problem is not that people cannot read detail, but they expect the detail to make sense.

Summary: **Picture styles**

- ***Use literal representation.***

  People who are unused to reading pictures will interpret the images very literally
- ***Avoid abstraction.***

  Abstract ideas are not readily understood from pictures
- ***Use three dimensions.***

  The representation of three dimensions in perspective is a powerful and relatively easily understood convention. But care should be taken to reinforce the three-dimensional effect with shadows, and recognizable objects, such as people to establish scale. There should be a viewpoint which allows every part to be seen clearly and unambiguously.
- ***Avoid distracting detail.***

  In drawings unnecessary detail should be avoided but relevant detail emphasized, and
- ***Do not assume that photographs are understood.***

  In photographs irrelevant detail should be omitted or physically cut out.

Symbols and conventions

The interpretation of any pictorial representation involves decoding a set of symbols. Always question what codes are being used and if your target audience shares the same code.

Symbols are abstract. They may, or may not, refer pictorially to the idea they represent. The skull and crossbones as a symbol of danger has connotations to something unpleasant. While the tick and the cross suggest good and bad. These symbols in themselves have no connection whatsoever to the ideas.

Symbols and conventions have to be explained. By themselves, they are usually not understood. But more than this, the idea of connecting a symbol to another object is unfamiliar to people unused to reading pictures. People who interpret pictures literally often see every object in the picture as separate, and assume no connection between them.

But among people with a common pictorial and conventional language, symbols and conventions are quick ways to communicate ideas. There are many things which cannot be represented graphically by literal representation.

(a) Good and bad

The tick and cross is the most common way of indicating what is good and what is bad. In many cultures, it is not a convention which is understood.

In Nepal, the tick and cross was used in a Unicef study to communicate the health message that breast-feeding was better than bottle-feeding (Figure 20). While a majority of the respondents were able to interpret the pictures literally as a woman breast-feeding and a woman bottle-feeding, only three per cent interpreted the tick and the cross as indicating good and bad.

In northern Pakistan, several attempts were made to convey the idea of a good and a bad house using various different conventions. In Figure 21, one-third of the respondents understood the tick and the cross as indicating good and bad. However, some of these did not associate the implied judgments of correct and incorrect to the two drawings. This means that even if people understand the symbols, they may not see the connection of the symbols to other objects. Other respondents described

Figure 20. In a study in Nepal only three per cent interpreted the tick and the cross as indicating good and bad ways to feed a baby.

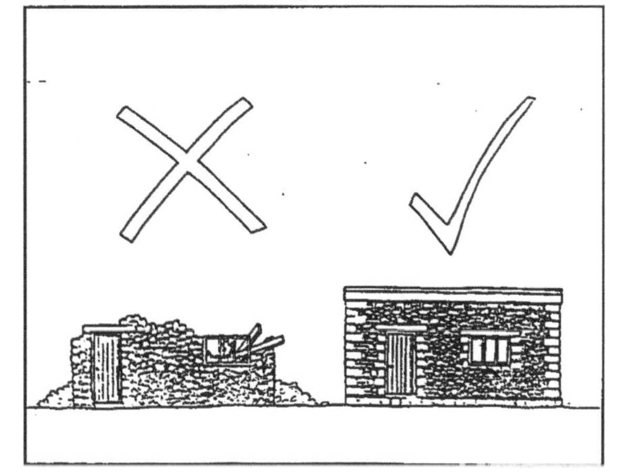

Figure 21. In this drawing indicating good and bad houses in northern Pakistan, the cross was interpreted as a ceiling fan and the tick as an Urdu 7.

the cross variously as a ceiling fan, the letter X, and an aeroplane. The tick was seen as a bent piece of metal or the Urdu number seven.

The picture was redrawn, using happy and sad faces instead of the tick and the cross (Figure 22). Some people recognized the facial expressions, but only a few connected the feelings expressed with the houses below. Those that did saw the faces as belonging to the owners of the houses, not as an abstract representation of the concept good and bad, indicating again the tendency towards a concrete interpretation of the image.

Figure 22. The use of happy and sad faces to indicate good and bad.

However, the fault lies partly with the drawing and the message. The two houses are not two comparable objects, and to say that a fallen-down or incomplete house is not as good as a complete one is not a useful message. A third version of the drawing was tried out, just showing the buildings. In many respects this drawing was better understood. People focused attention on the details of the buildings. They were not distracted by curious and irrational objects floating in the sky.

A potentially more useful drawing would have showed two complete houses, one well-built and one badly built and damaged but not collapsed. This would be a more realistic situation, and educationally more useful. It could prompt people to look for reasons why one house was damaged and the other not. This kind of analysis would usually need the intervention of a fieldworker or other educator.

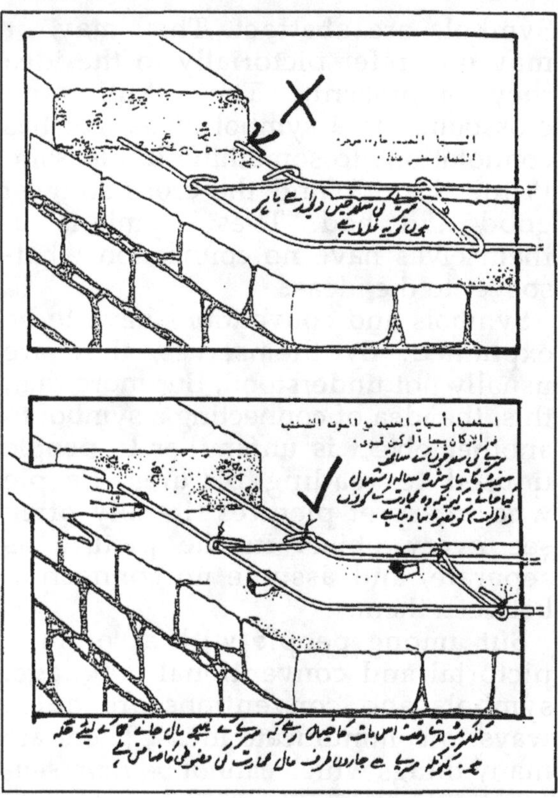

Figure 23. Drawing from a builder training manual in Yemen to show the correct placement of steel in a ring-beam.

Potentially, a better use of the tick and cross is an example from Yemen shown in Figure 23. It compares good and bad placement of steel within a ring-beam. The two drawings are comparing like with like.

When these drawings were tested in northern Pakistan, most people recognized the individual components of the drawing (stones and steel wire). But only one person understood the tick and the cross, and the message of the drawing. Several people picked out

Figure 24. Bad and good sanitation in Nepal.

that the steel in the lower so-called good drawing was tied together with wire. They could recognize something which was present in one drawing but absent in the other. But the subtlety of the location of the steel within the concrete was lost. The hand successfully served to provide scale to the drawing, but nobody realized that the two fingers were suggesting the distance of the steel from the base of the beam.

The borders around the images are an example of designer graphics which caused confusion. The confusion is strengthened by the hand and the steel passing over the border, reinforcing the impression that the border is some thing rather than simply the limit of the image. Some people suggested that the borders were rooms.

When the drawings were used in builder training courses in Yemen, no significant problems were experienced because the drawings were in the context of a discussion about the issue. In such situations, depicting the difference between good and bad practice is appropriate.

For audiences unused to reading pictures, the attitude is that something shown in a picture must be good. Otherwise, why would someone go to the trouble of making the drawing? Thus, in building education materials, showing bad practices should be avoided to the extent that it is possible. In materials designed for a training course, it is appropriate to make drawings of the bad as well as the good practices. These can be used in discussion.

Figure 25. Skull and crossbones in Somalia – danger or simply a man with large eyes?

Figure 26. Skull and crossbones in Nepal – one per cent saw danger, but sixty-one per cent saw something frightening.

In Nepal, frowning and smiling suns were used to indicate good and bad sanitation practices (Figure 24). No villagers interviewed understood the message, nor that the faces were expressing smile and frown, or good and bad. The concept of the picture simply made no sense to them. The idea that a non-human object, like the sun, can have an attribute is not understood. And, even if they understood this convention, they would not have made the connection between the sun and the action of defecating.

A standard symbol for depicting danger is the skull and crossbones. When testing the drawing in Figure 25 in Somalia, none of the respondents understood it conveyed the idea of danger. The most common response was that it was a man with large eyes.

Figure 26 is a simpler and less cluttered image which was tested in Nepal. Only one per cent said it indicated the intended abstract meaning of the symbol, that is, danger. However, sixty-one per cent saw it as something frightening or unpleasant, or to do with death. This result implies that the image has a good potential for being accepted as a symbol for danger, once explained properly to the audience. It is an image of death which rural people are especially familiar with. Thus there is a good basis for adapting it as a useful symbol.

(b) Movement

Illustrating the concept of movement inevitably involves using abstract conventions or symbols. Thus it is not likely to be understood without an intermediary to explain the idea. A common way to illustrate the idea is by movement lines. When interpreted literally, these lines make no sense.

Figure 27 was used in the Unicef study referred to above, which showed dramatic improvements in people's ability to interpret perspective drawings after six visits by educators. This picture of a man cutting wood was interpreted correctly by thirty-eight per cent of those interviewed. The study showed no significant improvement in people's skills in learning the concept of movement lines over the six visits.

Figure 27. Cutting wood in Nepal – understood by only thirty-eight per cent of the respondents.

Then, what is the difference? Why do people learn about perspective, but not about movement lines? In a perspective drawing, each object can be recognized literally, the objects make sense. Once the convention of perspective has been learnt, people can recognize the idea in real life – movement lines can never be seen since they do not exist.

Arguably, learning about the symbol of the tick and the cross involves even more of an abstraction, but the symbols are simple and standardized and most people learn to recognize them quite quickly. Movement lines are different from one drawing to another. To people who interpret pictures literally, this seems to be one of the most difficult concepts to get across.

Figure 28. Shaking a bottle of earth and water during an earth test. It seems unlikely that the target audience would understand the stylized representation of movement.

Figure 29. Dropping a ball of earth during an earth test. The movement lines in conjunction with a highly stylized representation are likely to confuse.

The drawings in Figure 28 and Figure 29, taken from two building education booklets from Ecuador, use movement lines to explain earth tests. It is likely that users of the booklets will experience similar problems of interpretation as those exposed to the man cutting wood in Nepal.

In Pakistan, a number of drawings were used to try and convey the idea of a house being shaken by an earthquake. The drawing in Figure 30 used movement lines around the top corners of the house and wavy lines for door and window frames, and for the ground.

The wavy lines of the frames and walls were interpreted as indications of a poorly built house. Some people went on to ascribe this to poor foundations due to bad ground as suggested by the wavy ground line. People readily commented on the bad window frames and the ground, because these were objects amenable to reason. The movement lines on top of the house were, however, largely ignored. When pressed to interpret, people either said they did not know, or suggested that water was running off the roof. Again, people try to make literal sense of the symbols or objects,

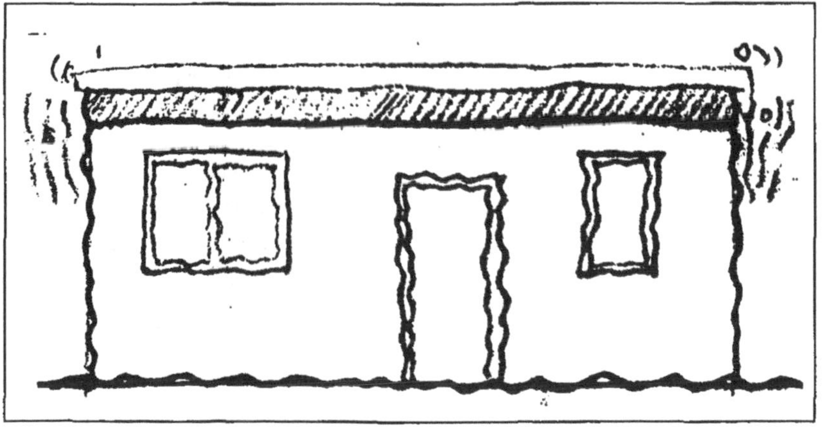
Figure 30. Movement in an earthquake, or a badly built house in the rain?

Case study: The sign of the cross in Ecuador

In March 1987 an earthquake struck a rural area of northern Ecuador. A number of locally based NGOs responded with aid programmes intended to help the villagers reconstruct their houses. Some of these programmes used the post-earthquake reconstruction as an opportunity to introduce small but sustainable improvements to the traditional forms of construction. It was hoped the techniques introduced would enter the vocabulary of construction of the region.

Most walls in the affected area are built of earth. Some are of mud bricks or wattle and daub. But most are built using rammed earth. In this technique, the wall is cast in place, rather like concrete. Since the cast earth block does not have to be moved, the timber mould can be much larger than those used to make mud bricks. The wall is cast in sections as the mould is moved along. When one layer has been done, other layers may be cast on top.

A local NGO called *Centro Andino de Accion Popular* (CAAP) chose to examine how the indigenous rammed earth technology could be modified to produce stronger corners. It was decided to experiment with a mould, L-shaped in plan, that can be used to cast the corner in a single piece. By making the short leg of the mould half the length of the long leg the mould can be flipped from one layer to the next to achieve joints which are as well staggered as in brickwork. Since, like traditional moulds, the mould is demountable it was possible to design a six-piece set which can also be used to make conventional straight sections.

The inside corner of the L was angled at 45 degrees. This made the initial carpentry rather more complicated. But it resulted in important structural benefits. First, the angle acts to distribute the forces induced in an earthquake which would otherwise be concentrated at the corner.

Second, the angle acts not only to distribute seismic forces but also those induced in the process of drying. Some neighbouring projects which subsequently adopted the L mould but without the angle frequently found that, in drying, the wall cracked diagonally across the corner. This lost the potential benefit of the L.

A problem which the prototype mould shared with traditional moulds was that the horizontal planks on the sides would mark the finished wall in a way which could obscure other horizontal lines caused through bad compaction or poor material. After discussion with the fieldworkers, who would have to provide future technical back-up to the programme, it was decided to make the sides of the mould with planks running vertically rather than horizontally. In this way any horizontal line seen on a wall could be readily identified as a problem by fieldworkers, builders, and householders alike.

The decision to have vertical planks resulted in a modified structural frame to support the planks. This purely structural necessity resulted in a striking cross-shaped structure. This was unique to the CAAP mould and could be identified from a great distance. The villagers across the region spontaneously named it the *tapialera de la cruz*, the mould of the cross.

In the implementation, an unexpected benefit emerged concerning the angle. In the finished building the angled corners are clearly visible from the inside. Whereas, a house built with non-angled L moulds is hard to distinguish from the traditional house built without the L.

Since all cultures are in some way status-conscious there is little incentive to use a new technology unless one can be seen to be using it. In the area of Ecuador affected by the earthquake the angle in the corner of the room has become an indicator of a well-built house and of a sensible householder.

The cultural value of both the cross-shaped structure and the angled corner as an index of sound construction were accidental spin-offs of the Ecuadorian experience. Afterwards, it could be seen that such characteristics which improve the power of the technology to communicate directly with the target audience could be planned for.

In the case of the angle it was realized that it would have been better still had there been a small angle on the outer corner as well. This would not have had a structural purpose, although it would have been useful in reducing accidental damage to the exposed and vulnerable corner. More important, it would have meant that the use of the technology would have been readily discernible, not just from the inside of the building but also from the outside. It would have had enhanced status value.

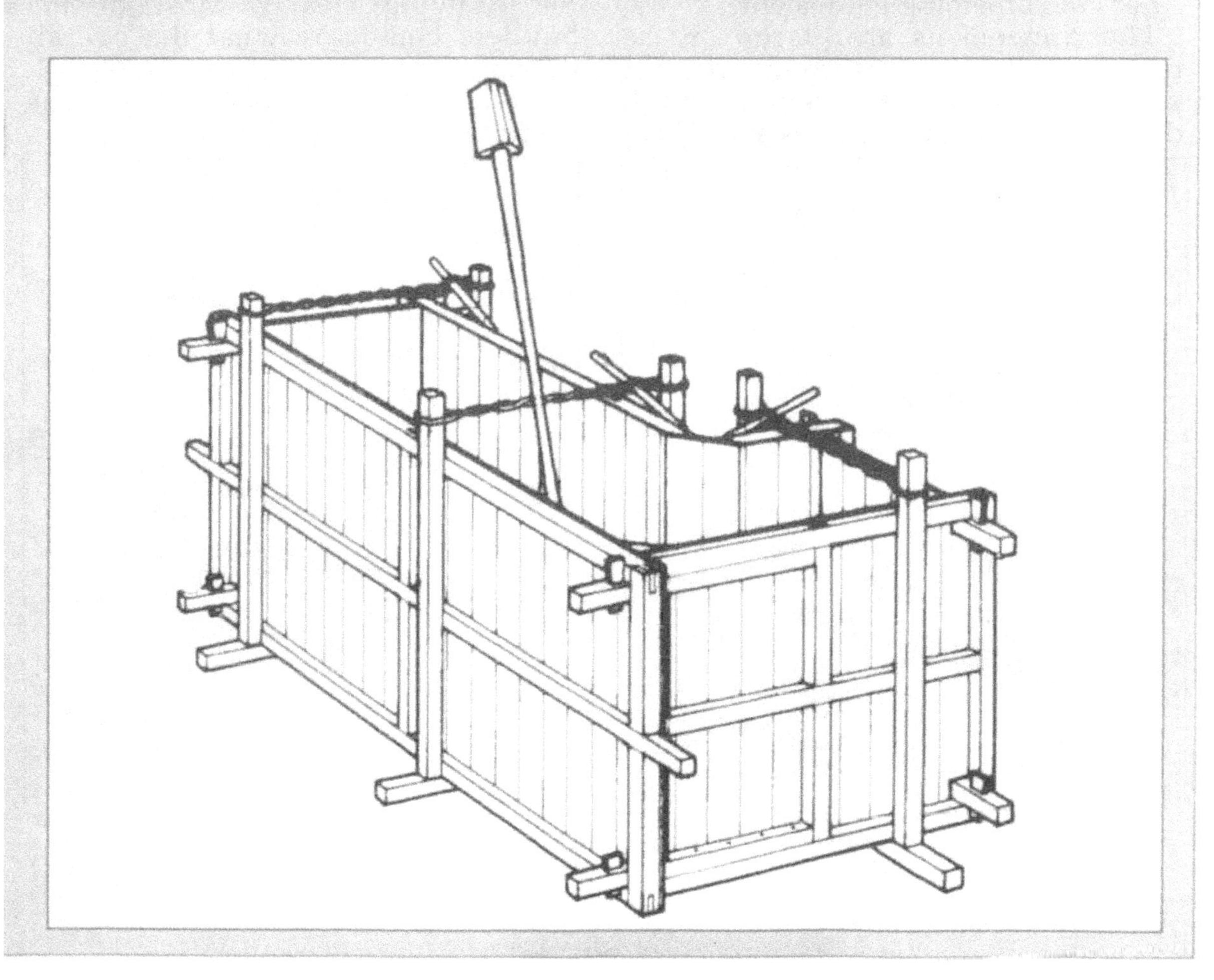

as these objects fit into their own concrete experience of life.

(c) Quantity

Builders need information indicating measurements, such as dimensions and quantities for various mixtures. The builders will usually be familiar with the nature of the tasks to be done, like mixing concrete, but may need to know how to adjust a mixture to make it stronger.

It is important that local forms of measurement are used, in the sizes of containers used and the form in which the materials are available on the market.

In Figure 31 a detail from a drawing of a concrete ring-beam is shown. The hand helps to indicate the scale, and the ruler introduces the idea of measurement – if the builders do use rulers or something equivalent.

The dimensions are placed on the picture following the common good practice of technical drawing, which dictates that the dimensions should not be on the object, but outside it. The idea is that the dimensioning lines should not obscure or be confused with the lines of the object itself.

Even though urban builders in northern Pakistan correctly read out the values of the dimensions, they consistently failed to understand what each dimension related to. Some confusion was also caused by the use of the ruler, because everyone knows that you do not measure buildings with a ruler but with a measuring tape. Thus, they tried to ascribe some other and more complicated meaning to the object, which diverted attention away from the central issue.

In a second version of the drawing, Figure 32, the dimensions were placed firmly on the object, and the dimension lines were made more bold. This made the drawing clearer. Most of the builders could say what the overall dimensions of the beam were, although some could not answer questions about the location of the steel.

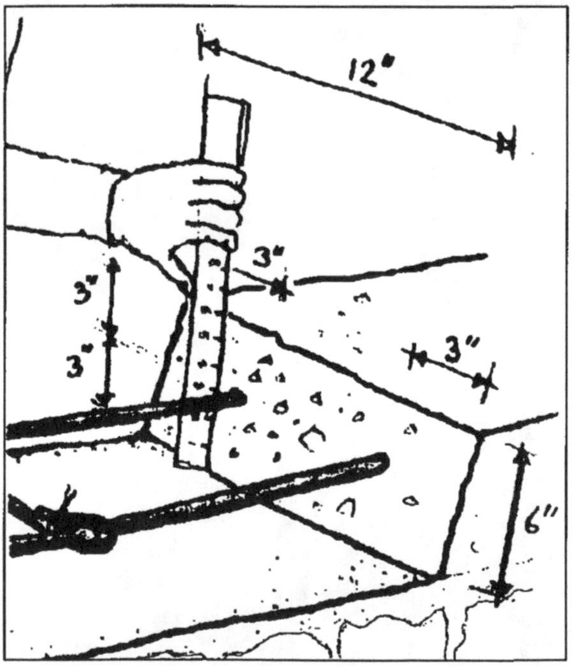

Figure 31. A detail of a concrete ring-beam. Placing the dimensions away from the object caused confusion.

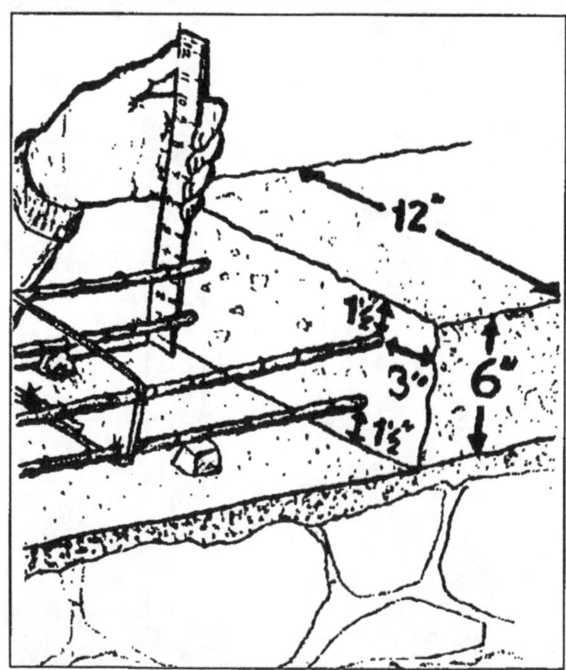

Figure 32. A detail of a concrete ring-beam. Placing the dimensions on the object helped interpretation.

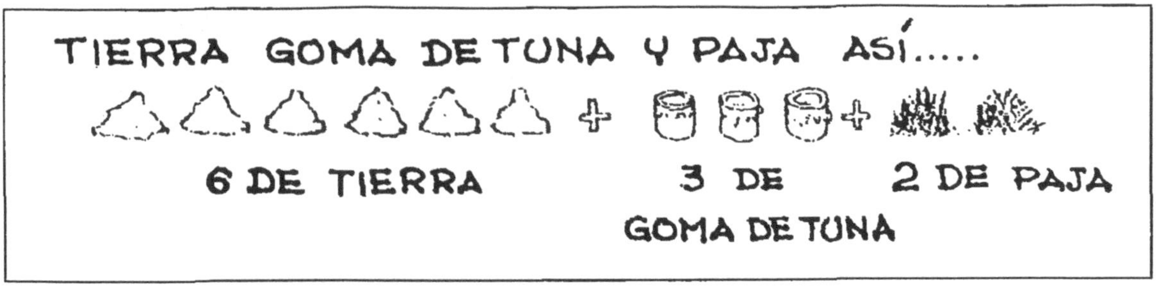

TIERRA GOMA DE TUNA Y PAJA ASÍ.....

6 DE TIERRA + 3 DE GOMA DE TUNA + 2 DE PAJA

Figure 33. This represents a mixture of earth, glue, and straw.

The concept of proportion in recipes for mixtures is difficult to illustrate. In Figure 33, a recipe for improved earth mixture is shown which illustrates some typical problems. There are problems relating to the idea of scale, the fact that building materials like cement, sand and earth are difficult to draw distinctly, and that abstract symbols like the plus sign may not be familiar to the reader.

When the graphic symbols being used are unclear, their repetition to indicate numbers is not adding anything. Also, many illiterate people are perfectly capable of reading numbers.

Figure 34 shows a picture used by Unicef in Nepal to describe making oral rehydration solution (ORS) from salt, sugar and water. A notable feature of the drawing is the fine and precise detail of the draughtsmanship. The glasses are shown clearly, the hand on the left adds a pinch of something, and the hand on the right gives an indication of quantity, but the picture on its own would not communicate the identity of the substance.

However, a good educational campaign does not depend on one piece of material alone to convey the message. It uses different materials through a variety of channels to reach several audiences in different ways. The Unicef campaign has put a great deal of effort into repeatedly designing and testing one memorable image, using it on posters, postcards, flip-charts and health centre curtains (see Nepal case study). The image as a whole has become a symbol for a message. Once people are familiar with how to make ORS, having learnt from a health worker or a neighbour, the symbol acts as a reminder of something which one already knows.

(e) Architectural conventions

Architects and engineers share an extended language of conventions which they may sometimes forget is not understood by others. Simple architectural plans are quite widely understood, because they relate to

Figure 34. A diagram used by Unicef in Nepal to illustrate mixing oral rehydration solution.

something which exists. When the lines of the foundation trenches of a house are laid out on the ground, people can see the plan. Similarly, the elevation of a house generally causes few problems because it is a picture of a house. In other words, these conventions lend themselves to literal interpretation, and are learnt relatively easily.

Figure 35 shows a cross-section of a house and its immediate context. This is a commonplace convention to the architect. But it can be mystifying to people unused to reading pictures, because it represents something which is never seen. We can also ask what is the reason for drawing a tree, or what looks like a tree, behind the house. This can be considered an unnecessary detail which can contribute to the confusion.

Architects and engineers also have methods to indicate certain materials. In Figure 35 the two-way hatching below the ground line is instantly recognizable to the architect as representing earth. Yet, to someone unfamiliar with the convention it is meaningless.

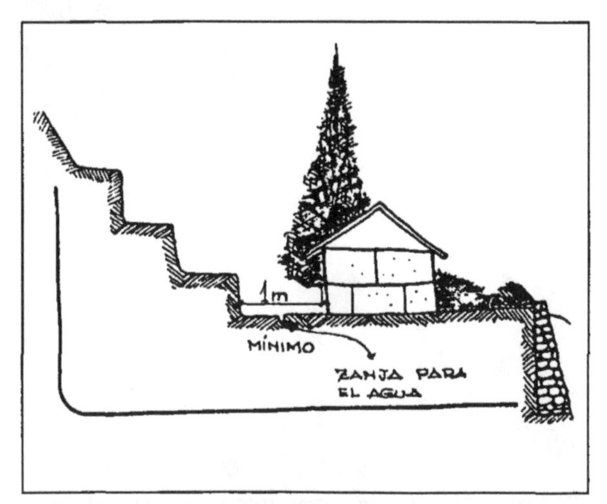

Figure 35. The use of a cross-section and other architectural conventions in a UN building education booklet.

Many of the conventions of architectural drawing relate to showing things which cannot otherwise be seen. Figure 36 from a builders' training manual, uses both dotted lines and transparency in an attempt to show how the roof timbers relate to the walls at the corners. This kind of complex three-dimensional detail is best suited to videos, models, and – best of all – demonstration on real buildings.

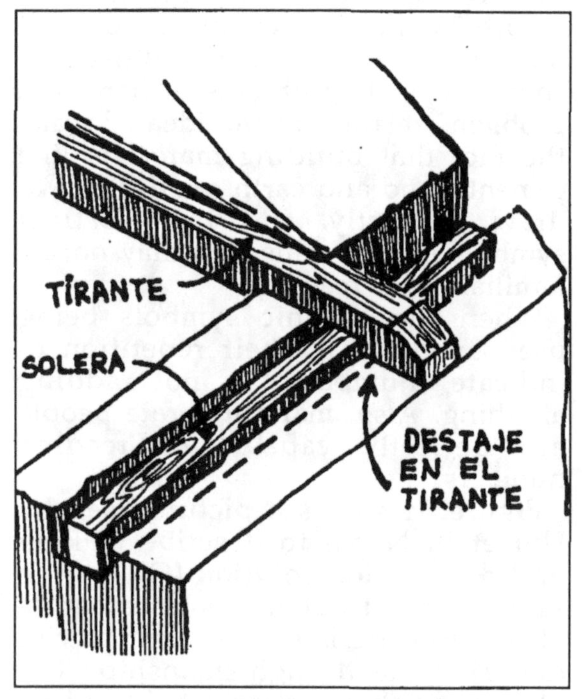

Figure 36. Detail from a post-earthquake reconstruction manual produced by an NGO in Ecuador. The use of both dotted lines and transparent walls to show what is beyond is potentially confusing.

Summary: Symbols and conventions

- ***Where possible, avoid symbols.***

  All communication with pictures requires some degree of interpreting symbols and conventions. If people are unfamiliar with a convention, they will try to read the picture literally. Thus, the use of conventions should be kept to a minimum in materials meant for people unused to reading pictures, especially if the materials are expected to work on their own

- ***Use common or realistic symbols.***

  If symbols have to be used, choose those that are as close as possible to the realistic idea one wants to communicate or which people are likely to come across elsewhere, such as the tick and cross

- ***Explain symbols and conventions.*** *When using symbols, one should take care to explain them at the bottom of the page, on a separate page at the beginning of the publication, or some other appropriate place, counting on the fact that there will usually be someone who can read and so help to interpret the symbols. This way, one will gradually build up people's skills in understanding symbols and conventions, and*

- ***Ensure than trainers understand the problem.***

  When symbols are used in training materials, a note should be made to the users about the difficulties in interpreting symbols, and suggest that the trainer takes time to explain carefully what each symbol means and how it is used.

Cartoons

For people already familiar with cartoon conventions, the cartoon is a potentially powerful, but also possibly patronizing medium, for conveying ideas rapidly. For those unfamiliar with the medium it is likely to be useless and confusing.

Cartoon conventions are frequently used in attempts to convey ideas about building improvement. Although the efficacy of this medium is uncertain it is known that:

- Cartoons are entertaining and very popular, especially with children in urban areas.
- Children and adults in rural areas in developing countries are usually unfamiliar with cartoons. But they can learn to read them if helped, and if abstract unfamiliar conventions like flash-backs and thought bubbles are avoided.
- People's tendency to interpret pictures literally needs to be modified if cartoons are to work as educational materials, and this may be a very difficult task.
- Cartoons can be good to dramatize a story, and to explain a sequence of events, such as what happens to a family with a poorly built or well-built house when a hurricane hits their village.
- Cartoons have been used in schools by many development projects to spread information on a variety of issues, such as sanitation and health. Studies have shown improvement in children's knowledge. But there has been no consistent evaluation which measures the effect of such cartoons on the children's behaviour.

What is not known is:

- To what degree cartoons are taken seriously as an educational medium by adults. Do adults – if they are familiar with cartoons as a medium – feel that the cartoon figures have got something to do with them, or do they see cartoons simply as for entertainment?
- To what degree do school children, who are the most common audience for cartoons from development projects, take the cartoons and the ideas home to discuss with their families?
- What level of literacy, both visual and verbal, is needed to really comprehend cartoons?
- Whether investing in the development of cartoons for adults not used to reading pictures is potentially an effective way of communicating ideas?

Cartoons have come to be widely used in building education materials. It is worth considering what assumptions the designers of these materials are making about the abilities of the target audience to decode the conventions of cartoons.

Figure 37. Happy house goes to Ecuador. Can an audience unfamiliar with drawings be expected to understand this level of abstraction?

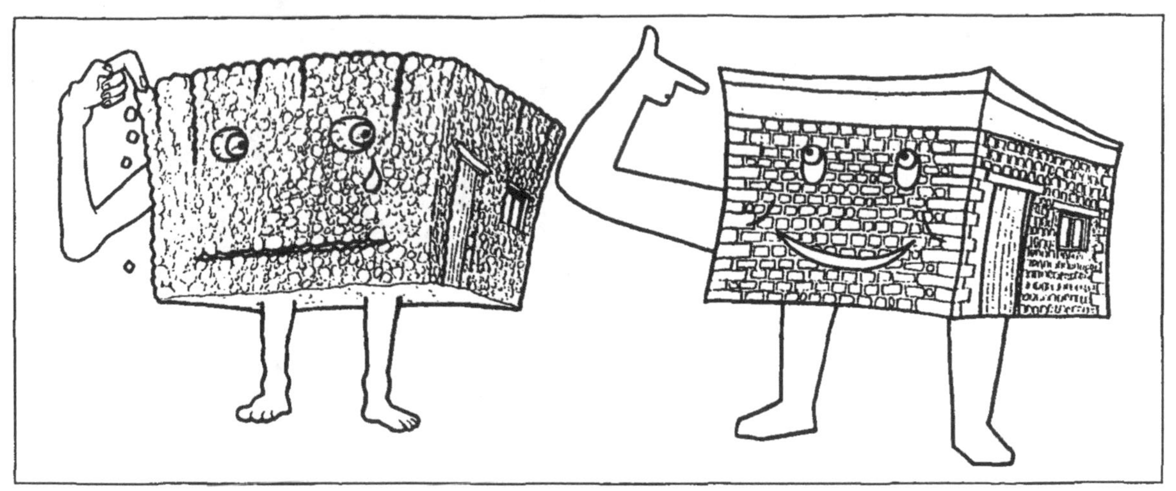

Figure 38. Happy house and sad house go to Pakistan. Or is one a house on stilts with a boat and an electricity meter on the wall and a staircase to the roof?

(a) Animating the inanimate

One of the most common practices in educational materials using cartoon figures is to give human characteristics to an inanimate object such as a house. This is called anthropomorphism. Experiments conducted in pictorial comprehension in Brazil concluded that giving human characteristics to animals was a problem for many people. Other studies have come to similar conclusions on a variety of subjects. It is probable that the same applies to buildings.

Figure 38 was produced to test the understanding of cartoon conventions among rural people in northern Pakistan. The happy house is well built with good stonework and a ring-beam to which the house is pointing. The so-called sad house, with poor rounded stonework and no ring-beam, is falling apart.

This drawing caused a great deal of confusion, with some people being visibly uncomfortable. Their anxiety seemed to come from correctly interpreting two quite contradictory sets of messages: features which said this is a person, and details which clearly belong to a house.

Some people simply gave up and said they could not understand it, while others struggled to make literal sense of the drawing. People described the eyes as electricity meters and the mouth of happy house as a boat or a window, while its arm was sometimes a staircase. One man said that there was a man trapped in the house, and only his eyes, mouth and feet were visible.

This is a good example of the principle that if a drawing as a whole does not make sense to people, they may pick out details and interpret them completely independently. What

Figure 39. Is this wall in Zimbabwe happy, sinister, or drunk?

would a boat be doing on the wall of a house?

A few younger people who had spent time in urban areas and seen films used the English word cartoon. They explained in great detail about the message of the drawing; sad house was sad because it was badly built and, in one case, happy house's firm legs were seen as indicating good foundations.

Figure 37 and Figure 39 from Ecuador and Zimbabwe show typical examples from building educational materials. Figure 40 is trying to give character and life not only to houses but also to the wind. People unfamiliar with cartoons and television commercials will most likely be highly confused by such pictures.

Figure 40. Drawings intended to suggest a good orientation for a house with respect to wind and sun.

(b) Bubbles

In cartoons, bubbles are used to indicate both speech and thought. A bubble with words or symbols is only comprehensible to those who can read. But even for the literate, the significance of speech and thought bubbles is not necessarily obvious.

For people who have learnt the basic convention of pointing, the stem of the bubble can seem to indicate pointing towards rather than coming from. When an Urdu version of Figure 41 was tested as a part of a larger cartoon strip, words such as poor old house were seen as captions pointing towards the speaker. The situation is confused since in some cases speech bubbles are indeed used to point towards objects, such as in Figure 11 where the bubbles are being used as captions.

In testing the cartoon story from which Figure 41 is taken the number of respondents was very small, and further work on the medium is required. But, in one interesting experience a literate man in the rural areas correctly read all the captions of the story but did not understand it at all. None of the funny captions made him laugh. The whole thing was so alien to him that he shook his head and did not know what to say about it. Cartoon humour is not necessarily transferable to another culture.

In Yemen, Figure 42 was used in an attempt to use thought bubbles to suggest that the owner of a house with a ring-beam can sleep soundly without fear of earthquakes. Here the use of the bubbles is ambiguous. It is not clear whether it is the house who is thinking or the owner. And, if the latter, why should he be thinking about himself? It was found that this and other cartoons were not widely understood.

Figure 41. Speech bubbles and animate houses, an earthquake, frost, and rain.

The use of bubbles with words requires basic literacy, and a certain agility of reading which enables the reader to dart about the image and piece the story together. To people of limited literacy, who may have to sound out every word and with limited exposure to pictures, the cartoon story and the message might not make sense.

Figure 42. In Yemen, thought bubbles to indicate peace of mind from having a sound house were not well understood.

(c) Strip cartoons

The strip cartoon is a communication tool demanding quite sophisticated skills from the reader. An understanding of cartoon conventions and the idea of sequence are required, both of which have proved to be problematic.

Figure 43 from Nepal is a particularly well-drawn strip cartoon without words and a minimum of conventions. Yet, only twenty-eight per cent of the respondents in a Unicef survey could correctly interpret it. Other strip cartoons currently in use in Nepal use many more cartoon conventions. They will most probably not be widely understood by rural people. In some circumstances the order of the sequence on the page may cause problems. When the Urdu version of the strip cartoon illustrated in Figure 41 was tested, some people read it from right to left in the same way that they read an Urdu script.

Figure 43. A Nepalese strip cartoon.

Figure 44. A more complex Nepalese cartoon using thought bubbles, alarm bubbles, movement lines, impact flashes, and an ambiguous order to the sequence.

(d) Photo-novelas

Photo-novelas are half way between real life and cartoons, that is to say cartoon techniques and ideas are used incorporating real people in real photographs. Photo-novelas are very popular in Latin America and elsewhere, notably India. One estimate claims that in Mexico some 70 million photo-novelas are sold every month. For many urban audiences in Latin America speech bubbles are easily understood. They are also regularly exposed to the codes of cartoon animation on television.

Studies have indicated that photo-novelas can be both popular and effective in promoting social change. They have great advantages over cartoons as they deal with real people. The disadvantage is they are more difficult and expensive to produce and they are mainly suited to an urban audience.

Figure 45. A Columbian photo-novela.

Summary: Cartoons

- ***Avoid being patronizing.***

  Even if an audience does understand cartoons, this does not necessarily mean that cartoons are a good way of conveying educational messages. Some people might feel patronized if presented serious information in a cartoon strip, as cartoons may be seen as a medium for children

- ***Assess people's understanding of cartoon conventions.***

  If a project decides to use cartoons, it should do so only after assessing people's perception of cartoon conventions. This would include such things as finding out about their opinions on animating the inanimate (anthropomorphism), and if they are familiar with speech bubbles and the like, and

- ***Field test everything.***

  Once the cartoons are developed, they need to be carefully tested for comprehension and acceptability.

Connections and sequence

Even when people understand individual images they can still have problems linking two or more images to make a single story. It is often preferable to have several incidents going on in one scene rather than a series of separate pictures. Where possible, only convey sequences in pictures when there is somebody there to explain what is going on.

We use a knowledge of graphic symbols to enable us to decode an image. We also use an understanding of various conventions which allows us to link several images together to tell a single tale. This may be in the form of a series of images in a video, pages in a flip-chart or booklet, or a set of images and text on a single page or poster.

Figure 46 was used in a sequence of drawings to try and convey the idea of an earthquake occurring. The human figure with a local-style hat flying off was included to try and stress that movement was actually happening at that moment. When the drawing was shown to villagers no connection was made between the figure and the state of the house. One person described the figure as a man dancing with a basket on his head. Making connections within a drawing is not an automatic skill. It is something which has to be learned.

Figure 47. When tested in Somalia, no one recognized the arrow as giving an indication of direction. The artist experimented and found that the pointing hand carried the idea much better.

(a) Pointing

One of the most basic ways of linking two images is the use of an arrow pointing from one to another. In the drawing from Yemen, shown in Figure 23, the arrow pointing to the incorrectly placed steel was not understood, while the hand which was there to indicate dimension was often wrongly interpreted as pointing to something. Similarly, in Figure 5, confusion was sometimes caused by the posture of the builder. The trowel in his hand, because it was very small, was read as a hand pointing to heaven. Some people thought that he was

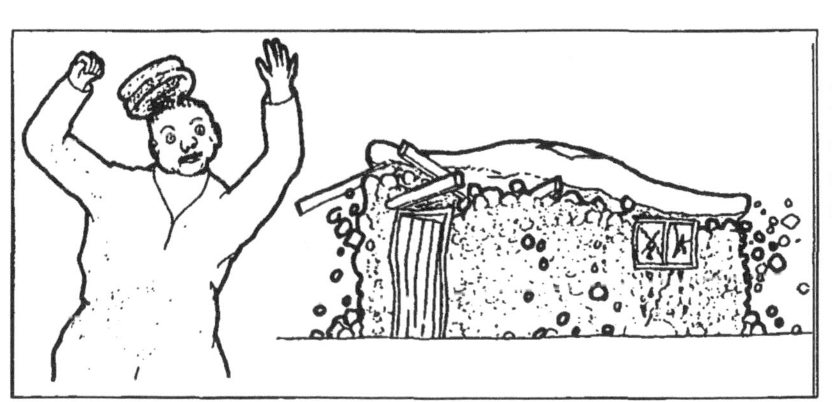

Figure 46. Does this represent an earthquake or a man dancing with a basket on his head?

praying. The pointing hand seems to be a powerful and widely understood symbol for pointing.

In Figure 48, which was tested with a small number of people in northern Pakistan, a pointing hand was used to link a detail with its context. Most people started interpreting the drawing of the house before moving to the detail above. At this point many respondents turned the drawing through 90 degrees, to make the disembodied pointing hand horizontal.

As they started to understand the drawing, they would turn it back to its intended orientation. However, more confusion would arise because the hand was felt to be pointing to some specific part of the detail drawing, such as the 2-inch dimension. To most respondents, the hand failed to suggest that the upper drawing was a detail of the lower drawing. The hand was not seen as the hand of the man illustrated, simply as a hand pointing to some aspect of the upper drawing.

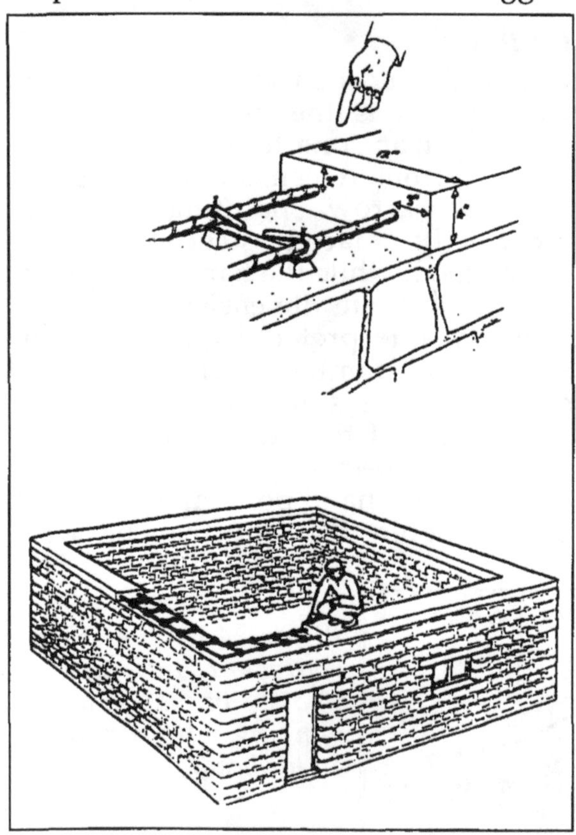

Figure 48. *Pointing to a detail. The hand in the detail was not seen as belonging to the man in the lower drawing.*

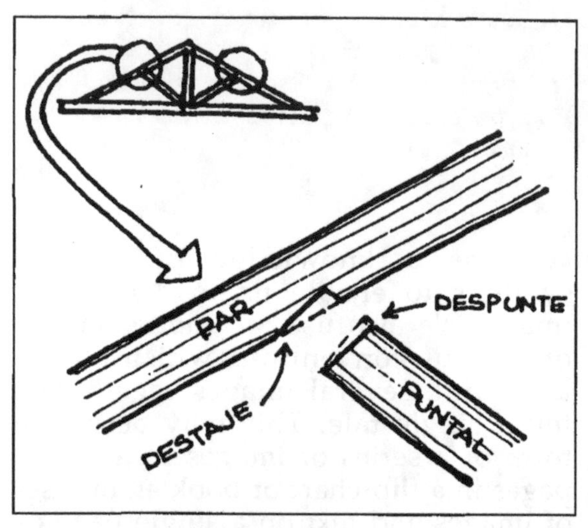

Figure 49. Highlighting a detail of a roof truss. Without explanation it is unlikely that the use of arrow, rings, and two different scales will be understood.

Various other experiments were tried in Pakistan which related a detail to a whole building using arrows, boxes and rings, and colour. None of these attempts was any more successful. In the example of a building education booklet reproduced in Figure 49, it is likely that many people would have problems relating the detail to the truss. The right hand circle, which does not have an arrow, is an additional potential cause of confusion.

However, in the Pakistan experiments when the connection was explained, many people understood what it was meant to be. Without the explanation, they could understand that there was a general theme about building a reinforced concrete beam, but not that one was a specific detail of the other.

If drawings such as Figure 48 were to be used in instructional building education materials, the man could be removed from the bottom drawing, and the hand could be drawn horizontal to point at the whole detail or omitted altogether, as it apparently does not do much to help the interpretation. In either case, such a drawing has to be pretested, and a note to the user about the difficulties in comprehending this convention has to be added.

One of the best techniques for showing the details of a building is to draw the building sufficiently large that the detail can be read on the drawing of the whole building. Figure 17 was one of the most successful of the drawings tested in Pakistan because it contained relevant, accurate, and familiar detail. Because the detail could be seen in context people could correctly interpret what was going on.

Figure 50. A ring-beam can protect your house in an earthquake.

(b) Cause and effect

Messages relating to cause and effect are commonly used in development programmes. For people unused to reading pictures, the connection is often not comprehended.

In disaster-related building education, there are two typical pairs of cause and effect which are often used:

- A natural disaster is happening or has happened. The effect is that houses are being damaged.
- A house has been built in a particular way. The effect is that the house has withstood the disaster.

The second of these is the key message which most disaster-related building education projects aim to convey. To convey that message requires constructing a series of ideas. If any link in the chain fails, the message is lost.

The drawing in Figure 50 was an attempt to convey the message that a ring-beam can protect a masonry house against an earthquake. During interviews in northern Pakistan, the houses were generally recognized as one house in good condition and the other as damaged. But the reason for the damage was unclear. When asked, most people ascribed the damage to rain. A few people saw the dark area on the front corner of the roof as indicating fire. After prompting, some people realized that there was a landslide on the mountain, but this was not seen as connected with an earthquake.

The wavy line in the foreground, meant to indicate movement, was totally perplexing to all respondents. In short, the drawing failed to establish even the first link in the chain of cause and effect. The approach which tried to get the audience to analyze various components of a single image and synthesize the elements into a single message seems to have been too ambitious. Similar problems with other cause and effect messages confirmed this finding.

(c) Pictures in a series

Sequences are often used in building education materials, particularly to illustrate the process of construction. The idea that pictures in a series, such as in a cartoon strip, are connected has to be learned. People unused to reading pictures will assume that each frame or each part of the picture represents a separate object or issue. This creates serious problems for communicating cause and effect, or a sequence of events.

A drawing is seen as portraying a static scene, not as part of a dynamic drama. Buildings portrayed as being damaged during an earthquake are seen as collapsed, incomplete, or under construction, not as collapsing.

Figure 52 was intended to convey the same message concerning earthquakes as Figure 50. Various clues were provided to make the sequence clear. Western-style numbers are widely understood in Pakistan by people who do not speak or read English. The use of bold numbers was intended to help lead people through the sequence. Since Urdu speakers read from the right, the layout was intended to suggest a movement from top right towards bottom left. Further clues were provided to convey the idea that the drawing showed four images of the same two houses. The ring-beam on the good house was coloured, while the door frame and the window frames of both houses were given different colours.

In practice, none of the respondents read the drawing as depicting two houses. Each of the eight houses were studied individually. The two right-hand houses at stages two and three were seen as poorly built houses rather than good houses successfully with-standing an earthquake. The wavy ground lines at stages two and three were sometimes read as indicative of poor foundations, occasionally as mountains, but most often they evoked no response unless a specific question was asked.

In a second version, from which Figure 46 is taken, the two men were given distinctive clothing in different colours, and were made larger with clear facial characteristics. Figure 46 corresponds with the left-hand house 3 in Figure 52. The man with a local-style hat flying off was included to stress that movement was happening at that moment. In Figure 52 it had been found that the collapsing houses were seen as static collapsed houses or houses under construction.

Figure 51. A comic strip without words which was field tested in Lesotho. Among the respondents who were illiterate, only sixteen per cent could fully understand the drawings. The layout of the sequence contributes to making it ambiguous.

Figure 52. A sequence intended to convey the experiences of a badly built and a well-built house before, during, and after an earthquake. Nobody understood it.

These changes made no difference to the recognition of the sequence, or suggested any movement. No connection was made between the figure and the state of the house. Each piece of the drawing was interpreted literally, in isolation from the other parts. Research in several other countries has also concluded that pictures in a series are not necessarily seen as being in connection with each other (see the Yemen case study).

Summary: Connections and sequence

- ***Explain connections between drawings.***

  Communicating a connection between two pictures, cause and effect, and a sequence of events to people unused to reading pictures will usually succeed only if there is a person to explain the issues. Even then, if the idea is completely new, people may take some time to learn. With repeated exposure to such pictures, they are likely to learn

- ***Do not assume that people make connections.***

  Before deciding to use pictures to portray connections between one or more things, assess the experience of the target audience in dealing with this convention, and

- ***Use connections and sequence in training courses.***

  Pictures in a series can be an effective way of communicating with builders in a training course, where discussion of problems and of cause and effect are part of the programme.

Cultural associations

The emphasis should not be on making images appear uniquely local but rather on making them not alien. Identify the codes of respectability of the audience. The use of religious symbolism should be handled with care.

When people can identify the objects and persons in a picture as being relevant to themselves and their situation, you have a good basis for communication. Educational materials need to take the local culture and customs of the target audience into account. This does not necessitate a long study by an anthropologist. It means the planners and designers have to undertake careful observation and interviews with local people before deciding on the contents and presentation of the educational messages.

The people in the picture do not have to be identical with the target audience, but the image portrayed must be recognized as a reachable or realistic goal for the audience. A picture of a movie star promoting better quality bricks might be admired by rural householders, but it will most likely not be effective as an educational message. The movie star is too different from the target audience. The respected local leader promoting the same brick, and living in a house built by such bricks, is a much more effective image.

Figure 53. A stylized house, but not to people whose houses have flat roofs.

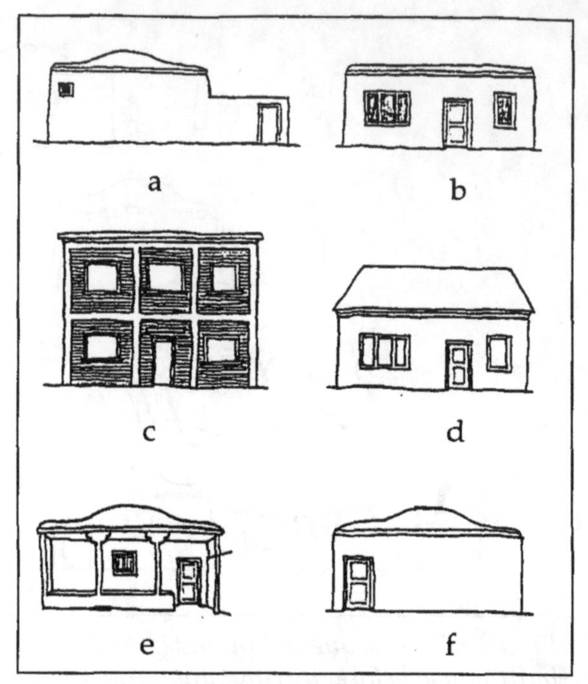

Figure 54. House types in the cold mountains of northern Pakistan. Which house is the warmest? For one man the answer was obvious – the rich man's, of course.

The drawing in Figure 53 is a simple but telling example. In Nepal, it was recognized by ninety-one per cent of the respondents in the eastern part of the country, but only by twenty-six per cent in the western part. The reason is that in western Nepal houses have flat roofs. At a simple level, taking the local culture into account can simply mean recognizing what the local physical environment looks like.

When Figure 54 was shown to villagers in northern Pakistan, nobody had any problem identifying the six images as houses. House *d* caused a few problems since its pitched roof was not clear. Some people saw it as a flat roof, others as a parapet.

Irrespective of this problem, house *d* with its large windows, like house *b*, was seen as belonging to the hot lowlands as opposed to the cold mountains where the interviews were being conducted.

Houses *a, e* and *f* were recognized as local houses because of the characteristic shallow dome to the roof. House *f* was the poor man's house because it was local and had no windows. While house *c* with two storeys and many windows was clearly recognized as the rich man's house. Quite simple drawings succeeded in communicating relatively detailed information about the location and quality of the houses. The drawings were based on knowledge of the country and observation in the field.

Beyond the first level of recognition of familiar objects, differences of cultural associations become less predictable. Nevertheless, there are some common issues.

(a) Unintentional subtlety

Though house *f* was widely seen as the poor man's house, it was also seen by many as the warmest house. Both interpretations were due to the lack of windows. But for one man, house *c* was seen as the warmest, because the rich man's house is always warm.

When tested on an urban audience, the rafters of collapsed houses in several drawings, like the bottom image in Figure 52, were widely seen either as missiles which had hit the house or as guns. Such ideas most probably come from exposure to television and movies, which the rural people do not have access to. Colourful posters depicting the most gruesome war scenes were also very popular in the town. Few people had any direct experience of war. In later drawings, such as Figure 21, the rafters were carefully drawn as rectangular, and the confusion with missiles and gun barrels was avoided.

Figure 55 was designed to test how a Pakistani audience would perceive a good builder. Man *a* was drawn to look like someone from the city. Man *d* was intended to look noticeably local with the characteristic hat, beard and waistcoat of the region. Man *b* and man *e* were meant to look like Pakistani working men, while man *c* was an attempt at a neutral figure with a minimum of distinctive features.

Man *a* was generally seen as alien to the building site. Man *d*, the local man,

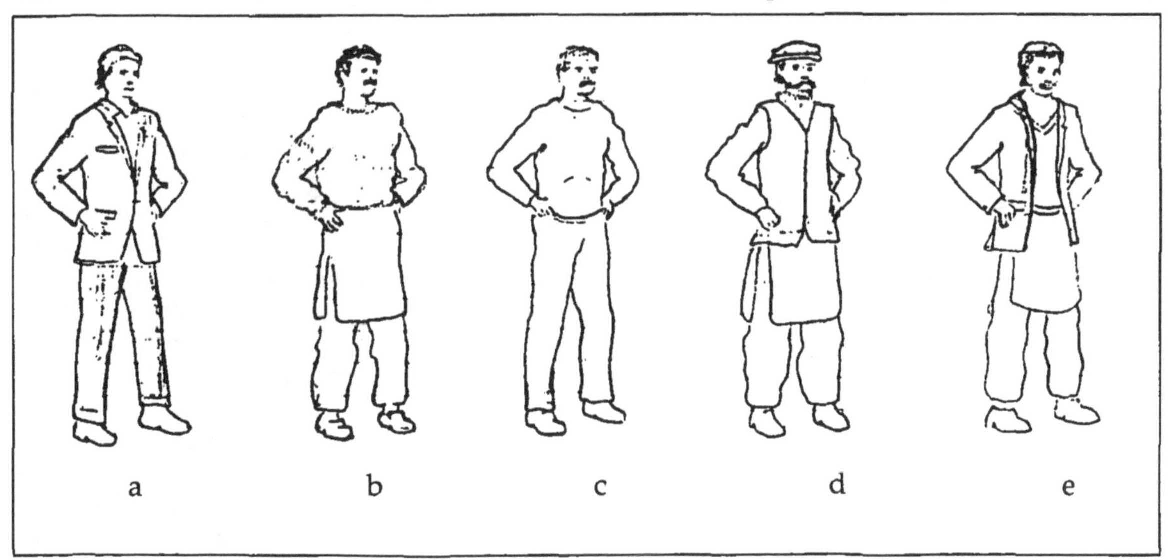

Figure 55. Differently dressed men in northern Pakistan. Who is the best builder?

provoked unexpectedly negative reactions. People saw subtleties in the hat which were both unintended and invisible to the non-local artist. The people interviewed were in the Hunza valley and their typical hat, though more or less of the same type, was slightly different to that of neighbouring Chilas. The hat was perceived as being from Chilas, and, due to the common rivalries of neighbouring peoples, the image provoked comments, such as: 'he won't be a good builder because he is a Chilasi'. Several people felt that he was old and thus would not be such a strong worker.

Men *b* and *e* were, in the eyes of the non-local artist, equally local and appropriate to the building site. Yet, man *b* was consistently picked out as the best builder of the five. When pressed for reasons it emerged that the subtle differences in the shape of the hands, the jaw and the chest conveyed the message that man *b* was stronger than the others. While the dress of *a* and *d* both conveyed negative messages, no features of the dress seemed to contribute to making any one figure more acceptable. Indeed, the one attempt to express localness, man *d*, largely backfired. The neutral figure, man *c*, provoked little reaction.

Although this picture was tested with more than 20 people, the answers are not conclusive. A preliminary conclusion would be that a picture of a local man should not be too distinctly local, but should include some elements of local dress which are common to a larger area. When testing several images, in this case six, it is wise to reduce the number gradually, say to three, once one is certain some images do not work. This will allow for more focused responses.

Figure 56 shows a colour postcard used in Ecuador to illustrate the use of limewash strengthened with glue as a finish to earth walls and floors. Care was taken to exclude extraneous detail from the photograph and the man is recognizably indigenous. Though his trousers, shirt and shoes are not uniquely local, neither are they at all unusual. However, prior to the photograph being taken the man had been working outside in the hot sun so he had put on the first hat he could find, which he is still wearing in the photograph. The hat is immediately recognizable to any local person as a traditional woman's hat. When the postcard was printed and used it was consistently found that people's attention would rapidly focus on the hat. A conversation would develop around the significance of a man wearing a woman's hat.

Figure 56. A postcard intended to illustrate improved limewash. In practice, the principal interest lay in why the man was wearing a woman's hat.

(b) Respectability of the role model
The key task in dissemination is to make the audience feel that people like you do this. In achieving this goal there is a need to ask to whom do people turn to for advice? Who influences them to make a decision to change certain aspects of their life? Who are their role models?

The respected people in the community will vary depending on the subject. The master builder and the local leader may be the ones community members seek out for advice on improving their houses, while the traditional midwife and the leader's wife may be the people women seek for advice on their children's health.

In printed materials, the same principles largely apply. In Nepal, health education teaching posters were being produced for use in the whole country. One topic was the need for supplementary feeding of children. The first version portrayed an urban mother with a child. Nepal is ninety-five per cent rural. When pretested, the village mothers did not relate to the urban role model. Her hairstyle, clothes and way of sitting were different, as were her baby's clothes and the manner in which she was feeding him using a spoon. The next version was a healthy, good-looking village mother, slightly better-off than the average villager. Her clothes were traditional, and she was feeding the baby with her hand the way all villagers do. This picture was widely accepted. It was an effective teaching tool because the audience could aspire to be like her, it was within their reach, and they could definitely copy what she was doing.

For a message to be considered for adoption, it should be passed through a credible source, someone the listener trusts, has respect for, and who seems relevant to the subject. Development programmes have experimented with

Figure 57. Gods used to promote hazard-resistant housing in west Bengal.

Case study: experimenting with gods in Nepal

In Nepal, the United Nations International Children's Emergency Fund (Unicef) has been experimenting with a wide variety of educational materials to promote the use of oral rehydration treatment (ORT). Apart from the more conventional techniques of posters, flip-charts, and videos they have experimented with such things as:

- Puppet shows
- Training retired Gurkha soldiers returning to their villages
- Merit badges for Scouts
- Radio shows and cassette tapes recorded by popular comedians
- Wall newspapers
- A comic strip magazine
- High-profile sporting events, and
- Cloth printed with ORT health messages and diagrams which is used as curtains in health centres and has been used for clothing.

An image of the Goddess Durga used on memory cards in an experiment to promote health care.

Their most ambitious experiment has been to try and enlist the help of traditional healers. There are some 600 000 traditional healers in Nepal, and most people, whether rural or urban, seek their advice. They are widely trusted, and always available in any village – which is not always the case with health workers. The healers are the most immediate and natural contact when anyone is ill.

Two early projects involved and trained healers, and also involved women in the villages who were already using ORT. The main medium was the village meeting where women discussed the pros and cons of medicine water. The traditional healers demonstrated its use and participated in the discussions. Without the benefit of any visual aids the use of ORT in the trial villages went up from about five per cent to more than seventy per cent in six months.

In an attempt to turn this local success into a nationwide phenomenon Unicef developed educational and promotional materials which try to harness the potential power of religious symbolism and associations. Apart from training, the healers are given a supply of small postcards, or memory cards. On one side of the postcard there is the standard image depicting the making of oral rehydration solution. On the other side, shown here, there is a full colour image of the Goddess Durga, who is the favourite god of the traditional healers. By associating the widely recognized image with Unicef's health message it is hoped that the message will be more acceptable both to the healers and to their patients. In addition, by printing attractive images on strong shiny cards it is

A strip cartoon in which the gods awake and travel around Nepal. They are shocked by the environmentally unsound development.

hoped that people will be more likely to keep and display the cards.

Unicef is also experimenting with more complex applications of the idea of linking gods with development messages, such as the one illustrated above. It remains to be seen whether the technique will work in the sophisticated and largely urban medium of the strip cartoon, and if people will accept the humanization of their gods.

However exciting the idea, it should be noted that the effect of this experiment is not yet known. The participation of the traditional healers and their acceptance of ORT has been the main factor in the acceptance of ORT by their neighbours. It is unclear to what extent, if any, the religious connotations of the memory cards have helped the process. Research in Nepal conducted in the 1970s that people did not link depictions of gods with other objects, the same way they did not link various objects in a picture to each other. Neither did they perceive that the picture was teaching them to take action.

It should be remembered that an intensive media campaign which successfully puts across new knowledge does not necessarily lead to the intended new actions. A survey in Nepal showed that after a long-term media campaign, forty per cent of the mothers interviewed knew about ORS, but only five per cent actually used it. A danger of high profile media campaigns is that they may overlook the difficult issues of what prevents traditional people from adopting and sustaining the new behaviour.

the application of this principle in a variety of ways.

Movie stars, sports champions, and local gods have been used by educators to link positive or negative figures to some action they want people to take or to avoid (see Figure 57 and the case study on Nepal). This can be seen as a culture-specific attempt to communicate good and bad.

Successful use of religious symbolism requires sensitivity to and in-depth knowledge of local beliefs and values. It also requires a careful tailoring of such messages to audiences who have the conceptual tools and experience to interpret them. Furthermore, the religion needs to be one which is amenable to be used this way without causing offence, and which is seen as an integral part of daily life.

In some cultures, there are religious beliefs associated with natural disasters. In India, images of Hindu gods have been used to depict the forces of nature, as in Figure 57. However, it is not recorded how these images are interpreted and regarded.

Famous individuals are often used to promote a product or an action. In many cultures it is common to find posters of political, religious and sporting leaders in the home. If the posters of development projects use the same sort of people, they too are liable to be widely displayed.

In Pakistan, the cricket star Imran Khan appears on health posters promoting vaccination. Questions can be asked about the potential effect of such an image:

- The image of a star personality will only be useful if his or her face is truly familiar to the target audience. This generally means that the audience has to be urban with regular access to television, the cinema and newspapers. In rural northern Pakistan where Imran Khan's poster was being displayed in health centres, few people recognized him. For this audience, the presence of the star could confuse rather than emphasize the main message of the poster, and
- The use of the star has to be relevant. Using a sports star to advertise milk as something he uses to keep healthy is relevant. One of the most well-known facts about Imran Khan is that, at the time of writing, he is unmarried. To promote vaccination for babies the use of a female star with a family would be more appropriate.

Local people who community members respect are usually the best role models. The model must be relevant to the subject. A role model for a message to rural villagers on improved houses for earthquake protection might be their local leader or master builder. It might be another respected person in the community who has applied the new technology to his or her own house.

Different types of leaders will be suitable to promote different themes. A religious leader or a medical doctor would most probably not be an effective role model for disaster-resistant construction methods. A farmer with well-kept fields, a tidy yard and a keen interest in new ideas might be a more relevant model.

The criteria and imagery which define respectability are very location-specific. But despite the world's rich cultural diversity there is a growing trend, even in remote communities, to equate modernity with respectability. When Figure 55 was tested, one man suggested that man a, the one in the suit, would be the best builder because he was a foreigner and so would obviously know best. Similarly, the ubiquitous man in a white coat is still a worldwide symbol for scientific expertise.

Figure 58. 'The composting toilet – the modern solution.'

(c) Culturally acceptable ideas

The language and the imagery used affect the way in which an idea is seen. An earth/cement block could be described as an improved earth block or as a low-cost concrete block. Which concept will be most attractive to the householder? A development institution concerned with using local materials may prefer to use an improved earth block. To villagers aspiring to modernity the idea of the low-cost concrete block may be more appealing.

The colour poster from Ecuador reproduced in Figure 58 promoted composting latrines as modern rather than appropriate or low-cost. By using bright colours on the building with white walls and bathroom tiles the intention was to promote a smart urban image. Also, a word equivalent to toilet was used rather than the word latrine, since latrines are what institutions give or recommend to the poor. Toilets and bathrooms are what people build from choice.

For three years prior to building and promoting the so-called modern solution the identical technology had been used in the same project with a more rustic image. Before the new promotion little interest had been shown in the latrine. The new urban-style bathroom immediately generated more interest.

In promoting an image of smart, modern respectability there is the possibility that the physical quality of the promotional material may affect the credibility of the message. To some audiences, a shiny printed flip-chart may give the young fieldworker a legitimacy which a hand drawn poster with the same content would not. On the other hand, the fieldworker's personal ability to communicate with the target audience and his or her credibility in the community will ultimately be more important than visual aids. Also, if the fieldworker has struggled to draw the posters personally, he or she may be much more motivated to use them well, and this will influence his or her personal communication.

(d) Use of colour

The use of colour is most effective when it is close to the natural and realistic colours people perceive in their environment. Unnatural colours used for effect can confuse the message for people unused to interpreting pictures.

In western cultures, the red and green of the traffic light is often used to signal danger or the coast is clear. In Islamic cultures, green is the colour for peace. In the Hindu culture, red is used for festivities and white for mourning. These colours by themselves will not promote the abstract meaning of peace or festivities, but they can be used to emphasize such issues. If a message in a Hindu culture is related to their festivals, giving the women red saris will give people a clue to the identification of festival time.

Colour can also confuse and destroy the message. In Nepal, the silhouette of a woman portrayed against the moon, taking a birth control pill, was intended to convey the idea of taking the pill at night. People interpreted the picture as if you take the pill, you will become black, which is a very negative connotation in a culture where black is seen as a colour for devils.

Summary: **Cultural associations**

- ***Find out how the audience see their problems.***

  Understanding cultural associations is extremely important when trying to communicate ideas to an audience. Finding out what is important to the audience regarding the actual issue or concept is the starting point

- ***Identify the codes of respectability of the audience.***

  Whom do they look up to, and what are their aspirations?

- ***Avoid things which are alien.***

  A drawing should be appropriate to the local environment. But avoid trying to be too clever in making people and objects too distinctly local. Rather ensure that nothing looks positively alien, and

- ***Adapt locally.***

  If all the householders in the country make up the target audience, the communicator has a major problem. In such cases, using just one image for all audiences may involve making so many compromises that the material could be close to useless. It will normally be necessary to prepare different material for, at the minimum, rural and urban audiences.

Use of text and language

For a picture of a technology to be memorable it must have a name associated with it. Text should be rigourously designed to be as clear as possible, rather than aesthetically pleasing. Where possible, text should make sense independently of the picture.

Short text clearly designed should accompany pictures used in promotional and educational materials. A good text will make it easier to remember the idea. People with limited literacy are often able to read short, clear texts.

(a) Describable ideas

The majority of houses in developing countries are built without the aid of drawings. Buildings are generally designed in discussion between a householder, members of his or her family, neighbours, and a builder. A builder can only be asked to build something which can be described in words and gestures, such as:

- A house five metres by seven, with a one metre deep veranda
- A tile roof sloping two ways
- An internal wall just here, and
- Door-frames like on my neighbour's house.

The role of educational material is to convey ideas, and also to make new ideas concrete enough for people to recognize them and say I want one of those.

One of the problems of introducing and communicating hazard-resistant construction is the abstract nature of

PUERTA PIVOTE

Hay varios problemas con las puertas que usan bisagras,

- *las bisagras son costosas.*
- *el marco tiende a separarse de la pared.*
- *la puerta tiene que ser bien liviana y resulta costosa.*

La forma tradicional de espiga de madera asentada en una piedra huequeada, se puede reproducir facilmente con dos pernos o chicotes de varilla.

Los pernos son fijados a la puerta y se pone en huecos hechos en los umbrales de arriba y de abajo.

La puerta no necesita marco y todo el peso de la puerta va directamente al piso. La puerta puede ser pesada y grande, y la luz y el aire pueden entrar por los dos lados del pivote.

Figure 59. A postcard promoting hanging doors on pivots. A single colour image, a short name, and a simple description combine to capture the idea.

some of the desirable technological practices. The idea of using the roof structure as a structural membrane to stiffen the whole house is important, but it is hard to give a name to it which captures the concept. In contrast, a ring-beam is an actual object which can be pointed at and named.

In northern Pakistan, many people use the English expression d.p.c. to describe the water-resisting layer of cement which ought to run around the top of the walls of a house to protect it from rain. Most do not know that the letters stand for damp-proof course. The acronym has become a name in itself for a recognized object. It may be that in this context the best way to promote a ring-beam is as an improved d.p.c. which protects the house against earthquakes as well as rain. This would fit with the existing vocabulary of architecture.

Figure 59 illustrates the combined use of text and colour photography on a postcard to promote a building technology. By putting a text label, plus a short one sentence description on the front of the card, people have a name to give to a definite idea. The back of the card states the problem which the technology addresses, and then describes the proposed solution.

(b) Clear text

Headlines and text illustrating pictures need to be clear and easy to read. Fashionable designer texts, such as that in Figure 60, are difficult to read, and should not be used in educational materials for people with low literacy levels. Figure 11 is another example of the kind of handwritten text which is commonly used. The low cross-bars to the A, E and H and the fat curves of the P and R obscure rather than assist legibility.

To the designer, the attraction of using only capital letters is that the resulting rectangular block of text looks neat on the page. But centuries of experience in calligraphy and typography suggest that text written in the Latin alphabet is clearest when:

- Both upper and lower cases are used. This makes it easier to recognize the beginning of sentences and the distinctive shapes of words
- Text characters are used which have serifs, which are the marks at the end of the strokes of many characters. This sentence with serifs is easier to read. This sentence without serifs is harder to read for people with a low level of literacy
- A font (style of text) is chosen which is both well-rounded and proportionally spaced, that is to say with less space for an i than for an m. Equally spaced text, such as that from traditional typewriters, can create the impression of gaps in the middle of some words, while others can become congested and confused. This is an example of a clear, good font, while this font can be more difficult to read
- Words and lines are amply separated to allow people to see each word clearly
- Columns of text are not too wide. One wide column in a text such as this makes it hard for the eye to find the beginning of the next line
- Words should not be split at the end of lines. For people struggling to make sense of words, breaking them up is just creating more problems
- Text is ranged lined up to the left margin only, like in this paragraph. It is common practice in books and newspapers to also line up the text on the right margin to form neat rectangular blocks of text. If the right margin is left ragged, the inexperienced reader is less likely to lose his or her place when scanning back to the beginning of the next line.

Figure 60. Sometimes, designer graphics impede rather than aid comunication.

These rules are well-known to calligraphers and typographers. They are sometimes set aside for aesthetic reasons. Breaking the rules may make the text more interesting for a highly literate audience. But, when communicating with people with limited reading skills, these rules are useful guidelines to more legible text. Similar kinds of good practice can probably be identified for non-Latin writing systems.

Rigorously applying such rules using either typeset text or a good quality computer printer will result in an apparently simple and formal layout. Clear, legible text is more important than attempts at being informal with handwritten text. Where mechanical lettering is not available, handwritten letters using well-rounded, upper and lower case, classic style is a good second best.

Figure 61. 'When the walls are ready they put the ring-beam on me.' Do adult builders expect houses to talk?

(c) Reading aloud

Even when the majority of the target audience is illiterate, there is still a place for text. In many circumstances there will be some people who can read aloud to others. To make reading aloud an effective start to the learning process, two conditions must be met. These are:

- There must not be too much text. One message explained in a short and simple way will keep the attention of the listeners and the interest of the reader. If it is too long it is unlikely that the reader will be prepared to read it aloud a second or third time, and
- The bulk of the text must make sense when read aloud without the pictures.

Captions to pictures in construction manuals are not always designed this way. When read aloud without the picture, they may not make sense. If a person is reading from a booklet to a small group, the group members will not be able to see the picture while listening to the text. Passing the booklet around will help, but will not solve the problem.

Particular problems can arise when the information is presented in cartoon form. If the text is in the form of speech, the listener has to understand who is talking. The problem is further aggravated when the speaker in the cartoon is the house, as in Figure 61. The result when spoken aloud is nonsensical and confusing.

Reading out the text is usually only the start of the communication process. Reading aloud can result in a discussion of the idea among the listeners and the reader, based on the text. Thus it is very important that the text is clear, to the point and carries the essence of the idea even without the picture.

Flip-charts are visual aids designed to show pictures to a group, with a trainer or teacher to explain the idea as an introduction to a discussion. The text is printed at the back of the picture, allowing the teacher to read while showing the picture to the audience. This principle can also be adapted to booklets which are designed for audiences with a low level of literacy by devices, such as printing the picture on one page and the text on the opposite page. In such cases an explanation to the reader should be provided on how to use the booklet in an effective way.

Development agencies often make educational materials in minority languages which are unique to a particular locality. Many of these languages have no history as written languages. When they are written, they are phonetic interpretations of the language in the script of one of the major international languages. Literate people from this group who can read such a text will almost certainly be able to read the international language, which will usually be the official national language of the country, as well. These publications only make sense if they are designed to be read aloud.

Figure 62. This drawing is suitable for a Quechua-reading person who understands cartoon conventions and hard to read so-called designer script, but does not understand Spanish. Thus it is designed for a person who does not as yet exist.

(d) Using computers

Computers and good quality printers have in recent years enabled many development institutions to produce their own finished text for educational materials. This technology is potentially a very good and powerful resource for the producers of educational materials.

Now it is possible for the prototypes of the materials developed for pretesting to be similar in quality to the finished product. This makes the results of the pretesting potentially even more useful. There are, however, several dangers in desk-top publishing which the development worker must take care to avoid:

- ***Intimidation.*** When people are faced with a rough draft, it is obvious that it is not in the finished stage. People are willing to comment and criticize because they believe their comments can still be used to improve the draft. When they are faced with a printed or finished-looking product, experience shows that people are much more reluctant to give their comments. The new situation poses a greater challenge to the pretester to convince the respondent that changes really are possible and easy in today's system, and that the comments will indeed be used
- ***Gimmicks.*** With the fascination of a new toy which can produce many elaborate and bizarre styles of text, some people are tempted into producing elaborate concoctions with half a dozen type styles in different sizes. This impedes, rather than assists, communication
- ***Too easy.*** Producing finished-looking materials becomes too quick and too easy. Previously, much thought would go into the design before it went for typesetting in the printing press. Now, the ease and cheapness

of production can reduce the pressure to get the design right

- ***Too many.*** Similarly, the ease of production combined with the high capital investment in the computer system can result in the production of too many ill-conceived and unnecessary materials, and

- ***Illusory quality.*** Almost anything printed on a modern printer gives the superficial impression of high quality. Inexperienced users of desk-top publishing programmes only use a fraction of the potential power which allows the fine tuning of text shape and spacing and which affects the legibility of the final result.

In other words, the danger is that the emphasis shifts to the techniques, and not on the assessment of the needs and aspirations of the target audience before materials are developed. As with other aspects of computers, the seductive quality of the latest equipment can lead one into thinking that lower quality equipment cannot do the job. This is often not the case. With suitable low-cost software the most basic printer, such as a nine-pin dot matrix printer, can be used to produce relatively high quality text, as in Figure 59.

Summary: Use of text and language

- ***Present ideas which can be given a name.***

  To adopt a new idea one must be able to put a name to it. In building education, the evolution of a visual vocabulary must develop hand in hand with an understanding of the local verbal vocabulary of building

- ***Design text to be read aloud.***

  As far as is possible, text should make sense when read aloud without drawings, and

- ***Test the drawings both with and without the text.***

  The link between words and images must be strong. Inappropriate text and drawings can cause confusion – is the problem with the drawing or with the text? To avoid such confusion, the drawing should first be tested on its own, and then with the text to assess exactly where the problem lies.

Summary: Illustrating building for safety

Picture style:

- ***Draw literally.*** People unused to reading pictures will interpret the images very literally
- ***Avoid abstraction.*** Abstract ideas in pictures can cause confusion
- ***Use three dimensions.*** Perspective drawings are relatively easily understood. The effect can be reinforced with shadows and objects, such as people, to establish scale and a clear viewpoint, and
- ***Stress relevant detail.*** In drawings unnecessary detail should be avoided. Relevant detail should be emphasized.

Symbols and conventions:

- ***Avoid unfamiliar conventions.*** If people are unfamiliar with a convention, they will try to read the picture literally, and
- ***Explain symbols.*** Symbols should be explained. This way, people's skills in.interpreting graphic conventions will be built up.

Cartoons:

- ***Only use if understood.*** The conventions of cartoons may be unfamiliar and unintelligible, and
- ***Avoid being patronizing.*** Cartoons may not be a good way of conveying educational messages. People can feel patronized if presented serious information in a cartoon strip.

Connections and sequence:

- ***Where possible, avoid them.*** Images are generally read individually
- ***Sequences can be useful when explained.*** Pictures in a series can be an effective way of communicating with builders in a training course, where discussion of problems and of cause and effect are part of the programme.

Cultural associations:

- ***Identify felt needs.*** Finding out what is important to the audience is the starting point
- ***Identify the codes of respectability.*** Who do people look up to, and what are their aspirations? And
- ***Avoid things which are alien.*** A drawing should be appropriate to the local environment. But, rather than trying to make people and objects distinctly local, ensure that nothing looks positively alien.

Use of text and language:

- ***Give ideas names.*** To adopt a new idea one must be able to put a name to it. The evolution of a visual vocabulary must develop hand in hand with an understanding of the local verbal vocabulary of building
- ***Use text which makes sense.*** As far as is possible, the text should make sense when read aloud, and
- ***Test drawings and words separately and together.*** Inappropriate text with a drawing can cause confusion. The drawing should be tested on its own first, and then with the text.

Preproduction testing:

- ***Always test.*** Many educational materials and techniques fail. New materials should be repeatedly tested with representative samples of the target audience, and
- ***Adapt locally.*** If one image is meant to serve for the whole country it will usually involve making so many compromises that it could be close to useless. It will normally be necessary to prepare different material for rural and urban audiences.

Conclusions:

- **Respect local knowledge and aspirations.**

  A successful building education project must build upon existing local knowledge and help to satisfy the aspirations of the target audience. To determine local knowledge and aspirations requires a dialogue based on mutual respect

- **Involve the beneficiaries at all stages.**

  Communication skills are required at all stages of a project and not just for the dissemination of how-to knowledge

- **Before trying to teach find out how people learn.**

  A building education project intended to introduce new building practices can best start by determining how new practices are already entering the society

- **Concentrate on one or two essential messages.**

  The development project which successfully introduces one wide-scale change of practice is doing better than most. The more messages that are promoted the less are the chances of any one of them being accepted

- **Identify clear targets and educational contexts.**

  No single educational tool, technique, or channel is going to be adequate for all audiences and tasks. All educational material should be designed with a particular audience and context in mind

- **Adapt educational techniques locally.**

  Educational tools and techniques must be tested and adapted locally. Although there are no universal solutions to communication tasks, there are universal problems

- **Use the real thing.**

  Demonstration buildings can be the most effective way of communicating improved construction, and

- **Invest in staff.**

  Staff at all levels need to be aware of the importance of communication skills. Investment in training of field staff in developing, testing, and using educational materials is vital.

Further reading

Social research

HASSOUNA, W. AND WARD, P., *Improving Environmental Health Conditions in Low-Income Settlements: a community-based approach to identifying needs and priorities*, 1987. World Health Organization, Distribution & Sales, 1211 Geneva 27, Switzerland, offset publication No.100, Order No. 1120100 US$9.60.

MCCRAKEN, J., PRETT, J., AND CONWAY, G., *An Introduction to Rapid Rural Appraisal for Agricultural Development*, 1988. International Institute for Environment and Development, 3 Endsleigh Street, London WC1H ODD, UK.

CHAMBERS, R., PACEY, A., AND THRUPP, L.A., *Farmer First: farmer innovation and agricultural research*, 1989. IT Publications, 103-105 Southampton Row, London, WC1B 4HH, UK, Order No. 011FFP, £3.15 + postage.

FEUERSTEIN, M-T., *Partners in Evaluation: evaluating development and community programmes with participants*, 1986. MacMillan Publishers, available from IT Publications, 103-105 Southampton Row, London, WC1B 4HH, Order No. 013PIE, £4.50 + postage.

CHAMBERS, R., *Rural Development: putting the last first*, 1983. Longman, available from IT Publications, 103-105 Southampton Row, London WC1B 4HH, UK, Order No. 013RDPTLF, £2.25 + postage.

Field testing pictures

HAALAND, A. AND FUSSELL, D., *Communicating with Pictures in Nepal*, 1976. Unicef, Lazimpath, PO Box 1187, Kathmandu, Nepal.

RANA, I., *Developing a Pictorial Language: an experience of field testing in rural Orissa*, 1992. IT Publications, 103-105 Southampton Row, London, WC1B 4HH, UK, Order No. 011DAPL, £6.95 + postage.

HAALAND, A., *Pretesting Communication Materials – a manual for trainers and supervisors*, 1984. Unicef, Burma.

MCBEAN, G., 9 *Rethinking Visual Literacy: helping pre-literates learn*, 1989. Unicef, Nepal.

Educational techniques

WERNER, D. AND BOWER, B., *Helping Health Workers Learn*, 1982. The Hesperian Foundation, PO Box 1692, Palo Alto, California 94302, USA, available from IT Publications, 103-105 Southampton Row, London WC1B 4HH, UK, Order No. 013HHWL, £10.95 + postage.

WILLIAMS, G., *All for Health*, 1989. Unicef, Facts for Life Unit, 3 UN Plaza, New York, NY 100017, USA, $1.00.

Printing processes

WILKINSON, J., *A Guide to Basic Print Production*, 1985. IT Publications, 103-105 Southampton Row, London WC1B 4HH, UK, in four volumes, Order No. 011GTBPPV1-4, £4.95 per volume + postage.

Picture credits

Report of a Field Trip to Northern Pakistan, 1991, by Andrew Coburn, Eric Dudley, and Ane Haaland. Figures 2, 3, 4, 5, 13, 14, 15, 17, 18, 21, 22, 30, 31, 32, 38, 41, 46, 48, 50, 52, 54, 55.

Construyamos en Tapial, 1987, Fundacion Ecuatoriana del Habitat, Ecuador. Figures 9, 11.

Communicating with Pictures in Nepal, 1976, Ane Haaland and Diana Fussell, Kathmandu. Frontispiece and Figures 1, 16, 20, 24, 53.

Construir la Casa Campesina, Centro Andino de Accion Popular, Ecuador, 1987. Figure 19.

Various booklets from project ECU-87-004 of UNCHS and the Junta Nacional de Vivienda del Ecuador. Figures 8, 29, 33, 35, 37, 40, 61, 62.

Understanding Print, Lesotho Distance Teaching Centre, Lesotho, 1976. Figures 7, 12, 51.

Rethinking Visual Literacy: helping pre-literates learn, George McBean, Unicef, Nepal, 1989. Figures 6, 27, 43.

La Casa de Tapial, Federacion Nacional de Organizaciones Campesinas, Ecuador, 1987. Figures 10, 28, 36, 49.

Field Trip Report from Somalia by Ane Haaland, 1986. Figures 25, 26, 47.

Various educational materials from the Dhamar Reconstruction Project, Yemen, 1984-6, by Jolyon Leslie. Figure 23 and Yemen case study.

Unicef Nepal. Figures 34, 42, 44, and Nepal case study.

Building Workbook for Secondary Schools, ZIMFEP, Zimbabwe, 1983. Figures 39, 60.

Various materials by Centro Sinchaguasin, Ecuador, 1984-8. Figures 45, 56, 58, 59.

Educational materials by UNNAYAN, Calcutta, India. Back cover and Figure 57.

Printed in the USA
CPSIA information can be obtained
at www.ICGtesting.com
JSHW060043150824
68134JS00028B/2611

9 781853 391835